# Fundamentals of Congenital Minimally Invasive Cardiac Surgery

WITHDRAWN FROM LIBRARY

D1339099

0968028

# Fundamentals of Congenital Minimally Invasive Cardiac Surgery

Edited by

**Vladimiro L. Vida**
**Giovanni Stellin**

University of Padua, Padua, Italy

Academic Press is an imprint of Elsevier
125 London Wall, London EC2Y 5AS, United Kingdom
525 B Street, Suite 1650, San Diego, CA 92101, United States
50 Hampshire Street, 5th Floor, Cambridge, MA 02139, United States
The Boulevard, Langford Lane, Kidlington, Oxford OX5 1GB, United Kingdom

Copyright © 2018 Elsevier Inc. All rights reserved.

No part of this publication may be reproduced or transmitted in any form or by any means, electronic or mechanical, including photo-copying, recording, or any information storage and retrieval system, without permission in writing from the publisher. Details on how to seek permission, further information about the Publisher's permissions policies and our arrangements with organizations such as the Copyright Clearance Center and the Copyright Licensing Agency, can be found at our website: www.elsevier.com/permissions.

This book and the individual contributions contained in it are protected under copyright by the Publisher (other than as may be noted herein).

**Notices**
Knowledge and best practice in this field are constantly changing. As new research and experience broaden our understanding, changes in research methods, professional practices, or medical treatment may become necessary.

Practitioners and researchers must always rely on their own experience and knowledge in evaluating and using any information, methods, compounds, or experiments described herein. In using such information or methods they should be mindful of their own safety and the safety of others, including parties for whom they have a professional responsibility.

To the fullest extent of the law, neither the Publisher nor the authors, contributors, or editors, assume any liability for any injury and/or damage to persons or property as a matter of products liability, negligence or otherwise, or from any use or operation of any methods, products, instructions, or ideas contained in the material herein.

**Library of Congress Cataloging-in-Publication Data**
A catalog record for this book is available from the Library of Congress

**British Library Cataloguing-in-Publication Data**
A catalogue record for this book is available from the British Library

ISBN: 978-0-12-811355-4

For information on all Academic Press publications visit our website at
https://www.elsevier.com/books-and-journals

Working together
to grow libraries in
developing countries

www.elsevier.com • www.bookaid.org

*Publisher:* John Fedor
*Senior Acquisition Editor:* Stacy Masucci
*Editorial Project Manager:* Sam Young
*Production Project Manager:* Mohana Natarajan
*Designer:* Christian Bilbow

Typeset by Thomson Digital

# Contents

# List of Contributors

**Lisa Ceccato** University of Padua, Padua, Italy

**Roberta Cabianca** University of Padua, Padua, Italy

**Annalisa Francescato** University of Padua, Padua, Italy

**Ana Pita-Fernández** Hospital Gregorio Marañón, Madrid, Spain

**Alvise Guariento** University of Padua, Padua, Italy

**Juan M. Gil-Jaurena** Hospital Gregorio Marañón, Madrid, Spain

**Maria T. González-López** Hospital Gregorio Marañón, Madrid, Spain

**Demetrio Pittarello** University of Padua, Padua, Italy

**Ramon Pérez-Caballero** Hospital Gregorio Marañón, Madrid, Spain

**Massimo A. Padalino** University of Padua, Padua, Italy

**Giovanni Stellin** University of Padua, Padua, Italy

**Karmi Shafer** University of Padua, Padua, Italy

**Chiara Tessari** University of Padua, Padua, Italy

**Vladimiro L. Vida** University of Padua, Padua, Italy

**Fabio Zanella** University of Padua, Padua, Italy

# Preface

The book "Fundamentals of Congenital Minimally Invasive Cardiac Surgery", edited by my friends and colleagues Drs. Giovanni Stellin and Vladimiro Vida, from the University of Padua, is an absolutely superb compendium of everything you need to know about starting and maintaining a minimally invasive practice in congenital and pediatric cardiac surgery. Minimally invasive pediatric cardiac surgery, which some think of as an oxymoron, is no such thing. I have always believed that it offers real value to the patient, be it cosmetic, psychological (no scar in the front), or biological (less bleeding, less pain, less sternum deformities without division of the manubrium). Of course, in this context, minimally invasive cardiac surgery is mostly about minimally invasive incisions, although less invasive approaches to cardiopulmonary bypass and other are also discussed. The book contains very detailed and graphic renderings of every step leading to a successful minimally invasive operation. The pictures are of exquisite quality, and the technical details are emphasized in a way that can only be done by surgeons with a vast experience in these types of surgeries. Indeed, approaching the heart from the side as in a mini-thoracotomy approach takes some getting used to, and the detailed tips about positioning, peripheral cardiopulmonary bypass, retraction, and instrumentation will help a lot of surgeons to start using this approach. In the end, practicing minimally invasive cardiac surgery is more about adopting a specific mindset and philosophy, and this book offers a superb platform on which to build and expand.

**Emile Bacha**
Chief, Cardiac, Thoracic and Vascular Surgery, Professor of Surgery, Columbia University,
NewYork-Presbyterian/Morgan Stanley Children's Hospital, New York, NY, United States

# Introduction

**Vladimiro L. Vida**

*University of Padua, Padua, Italy*

Improved surgical outcomes in patients with simple congenital heart diseases, combined with significant advantages in surgical and perfusion technologies, have stimulated surgeons to adopt minimally invasive technique both in adult and in pediatric patients. The aim of these techniques was to combine good functional outcomes with a reduction of surgical trauma for the patients and better final cosmetic result.

For 20 years we are treating with success in our institution simple and recently more complex CHD with the aid of minimally invasive surgical techniques. Since our initial experience there was a constant evolution of these techniques, with a progressive amelioration of initial technique and introduction of new techniques.

With this book we would like to provide practical guide to the most common used minimally invasive techniques for treating patients with congenital heart disease. A step-by-step illustrated detailed explanation of the surgical steps of the techniques together with illustrative videos is provided to facilitate the learning.

Chapter 1

# The Evolution of Minimally Invasive Cardiac Surgery for Patients With CHDs in our Institution

Giovanni Stellin

*University of Padua, Padua, Italy*

A routine median sternotomy has been the conventional approach for correction of congenital cardiac defects for many years. However, it often yields to poor cosmetic results with displeasure and psychological distress, especially in young female patients [1–5]. Surgery for sCHD has changed during the last decade, when different surgical techniques have been developed with the aim of combining good functional and cosmetic results.

Hagl et al. [6] in 2001 demonstrated that a full sternotomy is not always necessary for the correction of sCHD and other institutions have reported excellent results in the correction of sCHD by means of mini-sternotomy (MS) [6–11]. Furthermore, in other centers the use of a right anterolateral thoracotomy has been advocated [4] for repairing of simple and complex CHD [9–11].

Starting in the early 1990s, since August 1996, we have routinely adopted a systematic protocol of minimally invasive procedures for all patients with simple CHD, including perioperative 2D-echo monitoring (with both transesophageal and epicardial probes), postoperative pain control, early extubation, and early discharge from intensive care unit.

We arbitrarily have chosen different surgical approaches according to patient's age, gender, and specific patients' request, keeping in mind patient's satisfaction after the operation (what we called the "gender differentiated" minimally invasive surgery) [1].

A right anterior mini-thoracotomy (RAMT) (Image 1.1A and B) is less visible in females, especially for treating simple CHDs as ostium secundum atrial septal defects (ASD II), as the incision will remain within the submammary sulcus. A mid-line MS (Image 1.2) was offered as a surgical option mostly to male children, but also employed in females for repairing lesions other than ostium secundum ASD II due to a better exposure of the great vessels when other maneuvers are required (i.e., aortic cross-clamping, pulmonary valvotomy, closure of patent ductus arteriosus, etc.) [1].

The use of induced ventricular fibrillation (IVF) was adopted systematically in our Institution to avoid heart arrest, for short periods of cardio-pulmonary bypass. In fact, the use of short periods of IVF in association to the protective effect of a mild systemic hypothermia [9,12,13] has been proved to be a safe and effective strategy, when accurate intra-operative 2D-echo trans-esophageal monitoring for de-airing is employed before restitution of sinus rhythm

Since June 2006, as a refinement of our minimally invasive protocol, we have routinely employed peripheral cannulation (Image 1.3.) for initiating cardiopulmonary bypass in patients with simple CHD and a body weight superior to 5 kg and new retraction systems (Image 1.4). Both these technical applications allowed us a further miniaturization of the surgical accesses, thus allowing us to further reduce patient's surgical trauma (Images 1.5 and 1.6). We more recently added another minimally invasive surgical approach to our armamentarium, the right lateral mini-thoracotomy (Images 1.7 and 1.8) [14,15], which allowed us to widen the spectrum of treatable congenital heart defects.

Fundamentals of Congenital Minimally Invasive Cardiac Surgery. http://dx.doi.org/10.1016/B978-0-12-811355-4.00001-0
Copyright © 2018 Elsevier Inc. All rights reserved.

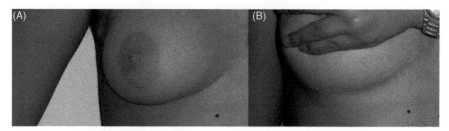

IMAGE 1.1 Postoperative image (A and B) showing the relationship of the right anterior minithoracotomy (RAMT) incision with the breast tissue in a female patient who underwent ostium secundum atrial septal defect (ASD) closure.

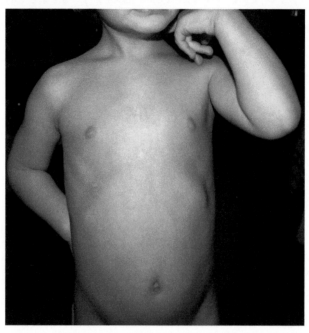

IMAGE 1.2 **Postoperative image showing a mid-line lower mini-sternotomy (MS) incision in a patient who underwent ASD closure (before 2007).**

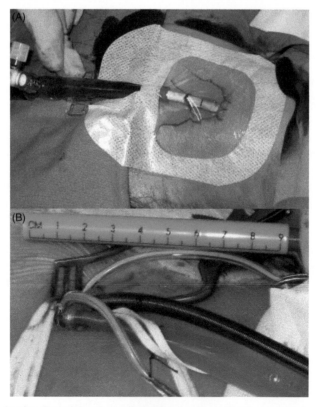

IMAGE 1.3 **Intraoperative image showing the peripheral cannulation for remote cardiopulmonary bypass.** (A) Percutaneous cannulation of the internal jugular vein, (B) surgical isolation and cannulation of the femoral artery and vein.

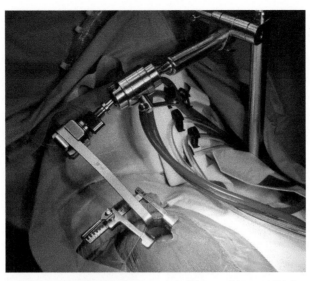

**IMAGE 1.4** Intraoperative image showing the use the Bookwalter retractor (Chapter 4) in a small infant (4-month-old, 5 kg patient) who underwent ventricular defect closure through a mid-line lower MS incision. Since 2007, we utilized a new retraction system (Bookwalter retractor), which facilitates the surgical exposure of the mediastinal structures further allowing us to minimize the extent of our surgical accesses.

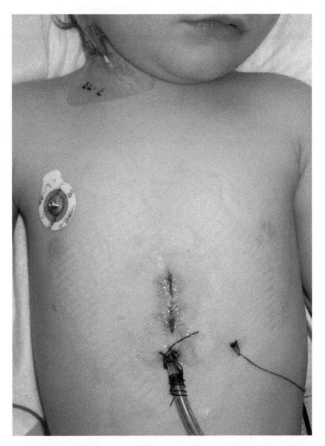

**IMAGE 1.5** Postoperative image showing the mid-line lower MS incision in a patient who underwent ventricular septal defect closure (after 2007).

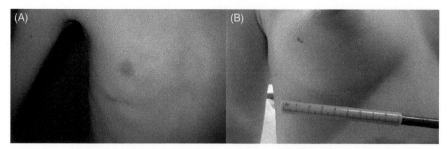

IMAGE 1.6 Postoperative image showing the evolution of the RAMT incision (A) in a prepubescent female patient and (B) in an adolescent female (with developed mammary gland and breast tissue). Both patients underwent an ostium secundum ASD closure. Since 2007, we minimize the size of RAMT incision moving it also more laterally (without the previous extension toward the mid line on the chest).

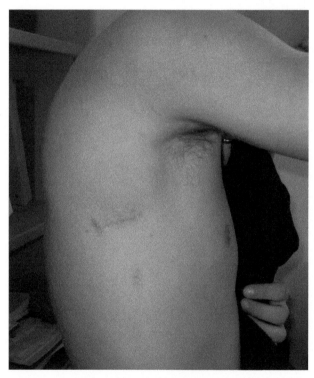

IMAGE 1.7 **Right posterior mini-thoracotomy (earlier in our experience with the development of this approach) (at the level of the angle of the right scapula), in a 16-year-old patient, who underwent the correction of a partial anomalous venous connection of the right lung.**

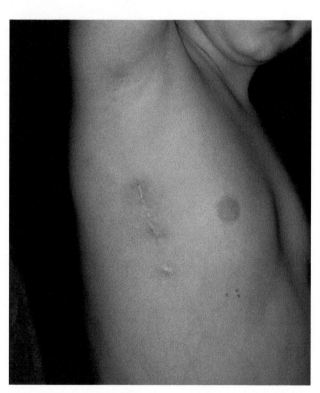

IMAGE 1.8 **Right lateral (axillary) mini-thoracotomy, in a 7-year-old patient, who underwent the correction of a partial anomalous venous connection of the right lung.**

# REFERENCES

[1] Vida VL, Padalino MA, Boccuzzo G, Veshti AA, Speggiorin S, Falasco G, et al. Minimally invasive operation for congenital heart disease: a sex-differentiated approach. J Thorac Cardiovasc Surg 2009;138:933–6.

[2] Umakanthan R, Petracek MR, Leacche M, Solenkova NV, Eagle SS, Thompson A, et al. Minimally invasive right lateral thoracotomy without aortic cross-clamping: an attractive alternative to repeat sternotomy for reoperative mitral valve surgery. J Heart Valve Dis 2010;19(2):236–43.

[3] Laussen PC, Bichell DP, McGowan FX, Zurakowski D, DeMaso DR, del Nido PJ. Postoperative recovery in children after minimum versus full length sternotomy. Ann Thorac Surg 2000;69:591–6.

[4] Ando M, Takahashi Y, Kikuchi T. Short operation time: an important element to reduce operative invasiveness in pediatric cardiac surgery. Ann Thorac Surg 2005;80:631–5.

[5] Dabritz S, Sachweh J, Walter M, Messmer BJ. Closure of atrial septal defects via a limited right anterolateral thoracotomy as a minimal invasive approach in female patients. Eur J Cardiothorac Surg 1999;15:18–23.

[6] Hagl C, Stock U, Haverich A. Steinhoff. Evaluation of different minimally invasive techniques in pediatric cardiac surgery. Is full sternotomy always a necessity? Chest 2001;119:622–7.

[7] Lancaster LL, Mavroudis C, Rees AH, Slater AD, Ganzel BL, Gray LA. Surgical approach to atrial septal defect in female. Right thoracotomy versus sternotomy. Am Surg 1990;56:218–22.

[8] Bleiziffer S, Schreber C, Burgkart R, Regenfelder F, Kostonly M, Libera P, Lange R. et al. The influence of right anterolateral thoracotomy in pre-pubescent female patients on late breast development and on the incidence of scoliosis. J Thorac Cardiovasc Surg 2004;127:1474–80.

[9] Vida VL, Padalino MA, Motta R, Stellin G. Minimally invasive surgical options in pediatric heart surgery. Expert Rev Cardiovasc Ther 2011;9:763–9.

[10] Abdel-Rahman U, Wimmer-Greinecker G, Matheis G, Klesius A, Seitz U, Hofstetter R, et al. Correction of simple congenital heart defects in infants and children through a minithoracotomy. Ann Thorac Surg 2001;72:1645–9.

[11] Massetti M, Babatasi G, Rossi A, Neri E, Bhoyroo S, Zitouni S, et al. Operation for atrial septal defect through a right anterolateral thoracotomy: current outcome. Ann Thorac Surg. 1996;62:1100–3.

[12] Cox JL, Anderson RW, Pass HI, Currie WD, Roe CR, Mikat E, Sabiston DC Jr. et al. The safety of induced ventricular fibrillation during cardiopulmonary bypass in nonhypertrophied hearts. J Thorac Cardiovasc Surg 1977;74(3):423–32. Sep.

[13] Vinas JF, Fewel JG, Arom KV, Trinkle JK, Grover FL. Effects of systemic hypothermia on myocardial metabolism and coronary blood flow in fibrillating heart. J Thorac Cardiovasc Surg 1979;77:900–7.

[14] Vida VL, Padalino MA, Boccuzzo G, Stellin G. Near-infrared spectroscopy for monitoring leg perfusion during minimally invasive surgery for patients with congenital heart defects. J Thorac Cardiovasc Surg 2012;143(3):756–7.

[15] Vida VL, Padalino MA, Bhattarai A, Stellin G. Right posterior-lateral mini-thoracotomy access for treating congenital heart disease. Ann Thorac Surg 2011;92(6):2278–80.

# Chapter 2

# Anesthesia for Minimally Invasive Cardiac Surgery

Demetrio Pittarello, Karmi Shafer
*University of Padua, Padua, Italy*

## EARLY EXTUBATION AND FAST-TRACK-MANAGEMENT

The postoperative care of patients with congenital heart disease (CHD) has changed over the years and in this context the early tracheal extubation after cardiac surgery is not a new concept.

In the literature, the definition of "early extubation" (EE) is not consistent and poorly characterized. Generally, the term EE is used when the endotracheal tube is removed within 6–8 h after the surgery or is associated with extubation in the operating room (OR).

Despite the age and complexity of the pediatric patients and the limited availability of drugs and ventilator technology EE was frequently performed in the 1960s and 1970s. Although the attempts to reduce the time of postoperative ventilation were successful, at the beginning, the technique never gained much popularity [1].

Now a days, EE is a part of the so-called "fast-track management," a term that refers to the concept of EE, mobilization and hospital discharge in an effort to reduce costs and perioperative morbidity. This approach has become increasingly popular [2–4], especially in mini-invasive surgery, with the delivery of cost-efficient care considered as an additional variable when measuring and comparing surgical outcomes [4].

The EE, albeit variably defined, has been implemented as a part of the management strategy for children undergoing cardiac surgery in mini-invasive surgery too [5]. This has been associated with improved resource usage by shortening the intensive care unit (ICU) and hospital lengths of stay in a range of patients with CHD [6,7]. Many of the early studies have demonstrated the feasibility of this approach in selected infants and older children.

There are many benefits to a fast-track approach to cardiac surgery. Potential advantages of EE includes: (1) less airway irritation and ventilator-associated complications (such as accidental extubation, laryngo-tracheal trauma, pulmonary hypertensive crisis during endotracheal tube suctioning, mucous plugging of endotracheal tubes, barotraumas secondary to positive airway pressure ventilation, and ventilator-associated pulmonary infections and atelectasis), (2) reduced parental stress, (3) reduced requirements of sedatives (and associated hemodynamic compromise), (4) more rapid patient mobilization, (5) earlier ICU discharge, (6) decreased length of hospital stay, and (7) reduced costs (ventilator-associated and length of ICU/hospital stay) [8–10].

Barash et al. claimed also psychological benefits of EE in addition to decreased pulmonary complications and duration of intensive care stay [11].

Despite this, the extubation in the OR after pediatric cardiac surgery is not a common practice. What are the fundamentals to move toward an attitude of EE?

Firstly, the migration to an EE-practice requires a shared mind-set among surgeons, anesthesiologists, intensivists, perfusionists, and nurses, where the majority of patients should be considered as potential candidates for extubation soon after surgery. After being said this, the principal characters that decide to attempt extubation in the OR should be the attending surgeon and anesthesiologist who were present during the surgical procedure. This allows for easy conversion to standard protocols, if indicated.

Fundamentals of Congenital Minimally Invasive Cardiac Surgery. http://dx.doi.org/10.1016/B978-0-12-811355-4.00002-2
Copyright © 2018 Elsevier Inc. All rights reserved.

**TABLE 2.1** Postoperative Opioid Analgesia

| Drug | Dose |
|---|---|
| Fentanyl | 0.5–1 mcg/kg/h i.v. |
| Morphine | 5–80 mcg/kg/h i.v. |
| | 5–20 mcg/kg/h i.v. neonates |
| | 200–500 mcg/kg/h oral 4 hourly |
| Remifentanil | 0.01–0.3 mcg/kg/min i.v. in ventilated patients |
| Codeine | 0.5–1 mg/kg oral 4–6 hourly |
| Tramadol | 50–100 mg i.v. 4–6 hourly (adult) |
| | 1–2 mg/kg i.v. 4–6 hourly |
| Sufentanil | 0.03–0.04 mcg/kg/h i.v. (adult) |

**TABLE 2.2** Postoperative Nonopioid Analgesia

| Drug | Dose |
|---|---|
| Paracetamol | 1 g i.v. 6 hourly (adult) |
| | 15 mg/kg oral 4 hourly (90 mg/kg max daily) |
| | 20 mg/kg rectal 6 hourly (90 mg/kg max daily) |
| Diclofenac | 1 mg/kg oral or rectal (3 mg/kg max daily)(adult) |
| Ibuprofen | 10–20 mg/kg oral or rectal (40 mg/kg max daily) |
| Ketamine | 10–45 mcg/kg/min i.v. |
| Clonidine | 2–5 mcg/kg oral 4 hourly |
| | 0.5–3 mcg/kg/h i.v. |
| Ketoprofen | 100 mg i.v. 12 hourly (adult) |

Secondly, the anesthesia technique is necessary for EE that needs to be a departure from the high-dose narcotic technique with postoperative ventilation and sedation and/or paralysis. To facilitate EE, others and we have used propofol, supplemented by low to moderate doses of narcotics. Moreover, analgesic drugs include opioids, local anesthetic agents, prostaglandin synthetase inhibitors (NSAID's), paracetamol, ketamine, and alpha-2 agonists (clonidine) (Tables 2.1 and 2.2). Opioids remain the primary analgesic agents after cardiac surgery because of their high efficacy. Co-analgesia with NSAID's and paracetamol plays a key role in reducing opioid requirements and side effects [12], which is particularly useful for fast-track surgery.

## INSTITUTIONAL MANAGEMENT AFTER MINIMALLY-INVASIVE CARDIAC SURGERY

After a favorable preliminary experience, our group implemented an institutional policy to attempt extubation in the OR in all suitable patients submitted to mini-invasive cardiac surgery, and the fast-track approach is applied to the whole of the patient's course from admission to discharge.

Anesthesia is not managed by a strictly defined protocol. However, our general approach for children considered to be fast-track consists of induction of anesthesia with thiopental 5 mg/kg, fentanyl 3 mcg/kg, and rocuronium 0.9 mg/kg. Throughout the procedure and during cardiopulmonary bypass (CPB), anesthesia is maintained with remifentanil 0.5–1 mcg/kg/min and propofol 3–5 mg/kg. Muscle relaxants are given again before CPB at 0.15 mg/kg. Intraoperatively, the decision of administering supplemental analgesia is left to the preference of the individual anesthetist.

Shortly before the end of surgery, remifentanil is discontinued and intravenous (i.v.) morphine (0.1–0.2 mg/kg) or fentanyl (1–2 mcg/kg) was administered.

After the end of the operation, the neuromuscular block is reversed by the i.v. administration of atropine 0.01 mg/kg and neostigmine 0.05 mg/kg. The adequacy of the reversal of the neuromuscular blockade is assessed by the head-lift test.

EE, either in the OR or in the ICU, is decided based on clinical evaluation and the following aspects are taken into consideration:

- Cardio-pulmonary bypass (CPB) time and aortic cross-clamp time, with a cut off at 150 min of CPB.
- Complexity of surgery.
- Hemodynamic stability and necessity of significant inotropic support, inotropic score (IS) < 5 or vasoactive-inotropic score (VIS) < 10.
- Bleeding, considered as clinically significant when it is observed a loss of 10%–20% of the circulating blood volume within the first 6 h after the surgery because its correlation with the need for transfusion of blood products and with the need of surgical re-intervention if it is not stopped [13,14].

OR extubation is never performed if there are signs of airway compromise, hemodynamic instability requiring bolus delivery of vasopressors, cardiac rhythm instability, excessive bleeding, or core temperature <35°C.

Eventually, final confirmation of the fast-track plan is made by the multidisciplinary team in the cardiac ICU, based on the surgical and anesthetic information at handover.

## MANAGEMENT AFTER VIDEO-ASSISTED THORACOSCOPIC SURGERY

In patients submitted to video-assisted thoracoscopic surgery (VATS), postoperative pain is significant and higher than surgical approaches via sternotomy, especially early after surgery [15,16]. Surprisingly, VATS is associated with an incidence of chronic pain similar to that of thoracotomy, with rates of pain ranging from 22% to 63% [17]. Chronic pain is relate to intercostal nerve and muscle damage with trocar insertion.

For these reasons, there has been increased interest in the use of paravertebral block (PVB) for VATS over the past decade. There is good evidence that PVB can provide acceptable pain relief compared with that provided by only infusional analgesia. The intercostal nerves are relatively devoid of covering fascia as they traverse the paravertebral space, making it an ideal location for local anesthetic blockade [18].

The traditional PVB technique is via a posterior approach using loss of resistance as the superior costotransverse ligament is traversed [19]. Recent modifications to this technique include use of a nerve stimulator [20,21] and ultrasound [22]. Alternatively, this technique can be performed by a single injection at the level of paravertebral space intraoperatively under direct vision by the surgeon [23] or anesthesiologists prior to chest closure [24]. In our experience, the PVB is performed by the surgeon under direct vision of the thoracoscopy before chest closure. The area at approximately the middle point of the intercostal space, 1 cm lateral to the sympathetic chain in the pleura, is located as the puncture site. Using a 24-G needle clamped by an oval forceps and attached via extension tubing to a syringe, the needle is advanced perpendicularly 0.5 cm beyond the pleura. After negative aspiration, 2-space injection PVB is performed by injecting of 0.5% ropivacaine (max 0.4 mL/kg) at the fourth and seventh intercostal spaces, with a 2-mL increment accompanied by the visualized slight distention of the pleura secondary to the spreading of the solution.

Important side effects such as hypotension, urinary retention, nausea, and vomiting are little frequent. PVB is associated with better pulmonary function and fewer pulmonary complications. Importantly, contraindications to other paravertebral approaches do not preclude PVB, which can also be safely performed in anesthetized patients without an apparent increased risk of neurological injury.

The single-shot multilevel PVB with ropivacaine 0.5% has a place in simple VATS procedures, longer and more complex procedures are well suited to PVB catheter insertion and infusion of local anesthetic [25].

This is the reason why the place of PVB in VATS surgery is clear, with analgesic benefits seen in the first few hours, and where a combination with an important postoperative analgesia component can reduce long-term adverse pain outcomes, although this is an appealing area for a well-planned large, prospective randomized trial.

At the end of the surgery whatever the anesthesia technique used, the extubation is performed when awake or while still deeply anesthetized at the anesthesiologist's discretion and patients are typically transitioned from facemask to nasal cannula oxygen in the OR.

## POSTEXTUBATION MANAGEMENT

Immediate postextubation analgesia or sedation in pediatric patients includes a single administration of rectal paracetamol (40–50 mg/kg), a bolus of morphine 0.05 mg/kg i.v. or fentanyl 0.5–1 mcg/kg i.v., and/or midazolam 0.05–1 mg/kg i.v., with repeated boluses as needed.

**TABLE 2.3 Inotrope Score and Vasoactive-Inotropic Score**

| |
|---|
| Inotrope Score (IS) [27] |
| Dopamine dose (mcg/kg/min) + Dobutamine dose (mcg/kg/min) + 100 × Epinephrine dose (mcg/kg/min) |
| Vasoactive-Inotropic Score (VIS) [28] |
| IS + 10 × Milrinone dose (mcg/kg/min) + 10,000 × Vasopressin dose (units/kg/min) + 100 × Norepinephrine dose (mcg/kg/min) |

**TABLE 2.4 Modality of Ropivacaine Administration at 0.2% (2 mg/mL) During Continuous Infusion in Pediatric Patients**

| Weight (kg) | mg/kg/h | Maximum hourly dose (mg) | Initial bolus (mL) | Continuous infusion (mL/h) | Duration (h) |
|---|---|---|---|---|---|
| ≤4 | <0.25 | 1 | 0.5 | 0.5 | Until 36 |
| 5–7 | <0.4 | 2 | 1 | 1 | 48 |
| 8–10 | <0.4 | 3–4 | 1.5 | 1.5 | 48 |
| 11–24 | <0.4 | 4.4–8 | 2–4 | 2–4 | 48 |
| ≥25 | <0.4 | From 10 | 5 | 5 | 48 |

For the first postoperative day in the majority of patients, morphine given by continuous i.v. infusion titrated to effect is sufficient for optimal pain control (usual dose: 10–20 µg/kg/h, following a 50 mcg/kg loading dose over the first hour) and fentanyl is used in place of morphine (usual dose: 0.5–1 µg/kg/h titrated to effect). When additional sedation or anxiolysis was required, an i.v. midazolam infusion (0.05–0.1 mg/kg/h) is administered.

Furthermore, rectal paracetamol is administered at 20 mg/kg 3 times a day for 3 days in patients with no oral fluid intake, and alternatively after the oral way is started, analgesia proceeded with oral or gastric administration of codeine, 1 mg/kg 3 times a day.

In adult patients postoperative analgesia consist of continuous infusion of sufentanil at the dose of 0.03 mcg/kg/h with an antiemetic agent as droperidol (2.5 mg) in an elastomeric pump for 2 days. In addition, paracetamol 1 g i.v. 3 or 4 times a day or ketoprofen 100 mg i.v. 2 times a day for 3 days.

Some authors advocate the use of local analgesia in addition to general anesthesia to facilitate EE [26]. Simple local anesthetic techniques can provide an important adjunct to any analgesia or sedation regimen, by reducing the background drug requirement and that for procedural analgesia. Local anesthesia can be given by a wide variety of routes that include topical, infiltration, peripheral nerve , or central continuous block (epidural or spinal). In our institution we used local infiltration with ropivacaine at the maximum dosage of 3 mg/kg (Table 2.3).

All patients receiving intravenous opioid analgesia are visited at least once daily by a member of the pain team (clinical nurse specialist and consultant anesthetist). Pain scores and vital signs are recorded by the bedside nurse, on an hourly basis throughout the day. In some cases when appropriate, pain scores are also recorded by the patient or parent. If the child is considered old enough to report his/her pain scores (using the 0–10 pain scale), these are also recorded. The decision to administer supplemental analgesia is made by the bedside nurse according to the pain scores (Table 2.4).

In conclusion, EE with good analgesia that allows rapid mobilization and recovery, is likely to be beneficial in terms of reducing postoperative complications in mini-invasive pediatric cardiac surgery.

In this kind of patients, a change in attitude of surgeons, anesthesiologists, perfusionists and nurses, combined with appropriate anesthetic and surgical techniques permitted better results and outcomes.

# REFERENCES

[1] Karski J. Practical aspects of early extubation in cardiac surgery. J Cardiothorac Vasc Anesth 1995;9:30–3.

[2] Lake CL. Fast tracking in paediatric cardiac anesthesia: an update. Ann Card Anaesth 2002;5(2):203–6.

[3] Laussen P, Roth S. Fast tracking: efficiently and safely moving patients through the intensive care unit. Prog Pediatr Cardiol 2003;18:149–58.

[4] Metin K, Celik M, Oto O. Fast-track anesthesia in cardiac surgery for noncomplex congenital cardiac anomalies. Pediatr Neonatol 2005;5:2.

[5] Alghamdi AA, Singh SK, Hamilton BCS, Yadava M, Holtby H, Van Arsdell GS, et al. Early extubation after pediatric cardiac surgery: systematic review, meta-analysis, and evidence-based recommendations. J Card Surg 2010;25:586–95.

[6] Lawrence EJ, Nguyen K, Morris SA, Hollinger I, Graham DA, Jenkins KJ, et al. Economic and safety implications of introducing fast tracking in congenital heart surgery. Circ Cardiovasc Qual Outcomes 2013;6:201–7.

[7] Howard F, Brown KL, Garside V, Walker I, Elliott MJ. Fast-track paediatric cardiac surgery: the feasibility and benefits of a protocol for uncomplicated cases. Eur J Cardiothorac Surg 2010;37:193–6.

[8] Alexander J, Mittnacht C, Thanjan M, Srivastava S, Joashi U, Bodian C, et al. Extubation in the operating room after congenital heart surgery in children. J Thorac Cardiovasc Surg 2008;136:88–93.

[9] Heinle JS, Diaz LK, Fox LS. Early extubation after cardiac operations in neonates and young infants. J Thorac Cardiovasc Surg 1997;114(3):413–8.

[10] Kloth RL, Baum V. Very early extubation in children after cardiac surgery. Crit Care Med 2002;30(4):787–91.

[11] Barash PJ, Lescovich F, Katz JD, Talner NS, Stansel HC Jr. Early extubation following pediatric cardiothoracic operations: a viable alternative. Ann Thorac Surg 1990;29:228–33.

[12] Hiller A, Meretoja OA, Piiparinen S, et al. The analgesic efficacy of acetaminophen, ketoprofen, or their combination for pediatric surgical patients having soft tissue or orthopedic procedures. Anesth Analg 2006;102:1365–71.

[13] Williams GD, Bratton SL, Ramamoorthy C. Factors associated with blood loss and blood product transfusions: a multivariate analysis in children after open-heart surgery. Anesth Analg 1999;89(1):57–64.

[14] Savan V, Willems A, Faraoni D, Van der Linden P. Multivariate model for predicting postoperative blood loss in children undergoing cardiac surgery: a preliminary study. Br J Anaesth 2014;112(4):708–14.

[15] Vogt A, Stieger DS, Theurillat C, Curatolo M. Single-injection thoracic paravertebral block for postoperative pain treatment after thoracoscopic surgery. Br J Anaesth 2005;95:816–21.

[16] Nagahiro I, Andou A, Aoe M, et al. Pulmonary function, postoperative pain, and serum cytokine level after lobectomy: a comparison of VATS and conventional procedure. Ann Thorac Surg 2001;72:362–5.

[17] Landreneau RJ, Mack MJ, Hazelrigg SR, et al. Prevalence of chronic pain after pulmonary resection by thoracotomy or video-assisted thoracic surgery. J Thorac Cardiovasc Surg 1994;107:1079–85. discussion 1085–1086.

[18] Bertrand PC, Regnard JF, Spaggiari L, et al. Immediate and long-term results after surgical treatment of primary spontaneous pneumothorax by VATS. Ann Thorac Surg 1996;61:1641–5.

[19] Karmakar MK. Thoracic paravertebral block. Anesthesiology 2001;95:771–80.

[20] Eason MJ, Wyatt R. Paravertebral thoracic block: a reappraisal. Anaesthesia 1979;34:638–42.

[21] Lonnqvist PA, Olsson GL. Paravertebral vs. epidural block in children. Effects on postoperative morphine requirement after renal surgery. Acta Anaesthesiol Scand 1994;38:346–9.

[22] Jamieson BD, Mariano ER. Thoracic and lumbar paravertebral blocks for outpatient lithotripsy. J Clin Anesth 2007;19:149–51.

[23] Sabanathan S, Smith PJ, Pradhan GN, et al. Continuous intercostal nerve block for pain relief after thoracotomy. Ann Thorac Surg 1988;46:425–6.

[24] MylesPS, Bain C. Underutilization of paravertebral block in thoracic surgery. J Cardiothorac Vasc Anesth 2006;20:635–8.

[25] Cereda CM, Brunetto GB, de Araujo DR, de Paula E. Liposomal formulations of prilocaine, lidocaine and mepivacaine prolong analgesic duration. Can J Anaesth 2006;53:1092–7.

[26] Hammer GB, Ngo K, Macario A. A retrospective examination of regional plus general anesthesia in children undergoing open heart surgery. Anesth Analg 2000;90:1020–4.

[27] Wernovsky G, Wypij D, Jonas RA, Mayer JE Jr, Hanley FL, Hickey PR, et al. Postoperative course and hemodynamic profile after the arterial switch operation in neonates and infants. A comparison of low-flow cardiopulmonary bypass and circulatory arrest. Circulation 1995;92(8):2226–35.

[28] Gaies MG, Gurney JG, Yen AH, Napoli ML, Gajarski RJ, Ohye RG, et al. Vasoactive-inotropic score as a predictor of morbidity and mortality in infants after cardiopulmonary bypass. Pediatr Crit Care Med 2010;11(2):234–8.

# Chapter 3

# Cardiopulmonary Bypass Strategies: Vacuum Assisted Venous Drainage

Fabio Zanella, Lisa Ceccato, Vladimiro L. Vida

*University of Padua, Padua, Italy*

## VACUUM ASSISTED VENOUS DRAINAGE

Vacuum-assisted venous drainage (VAVD) (Figs. 3.1 and 3.2) is a widely used technique that cardiac surgeons and the perfusion team components use to reduce hemodilution [1–5] and to reduce the size of the cannula during minimally invasive cardiac surgery (MICS) [4,6,7]. Such assisted venous drainage allows to decrease the declivity of the reservoir, and thus, to significantly decrease the length of the venous, arterial, and suction lines. The result is a shorter and smaller tubing system with a reduced priming volume [2]. Furthermore, VAVD has proven to increase the venous return through cannula that demonstrate limited flow capacity under siphon drainage conditions [8].

During MICS, surgeons perform surgical procedures through small incisions. To allow this approach, VAVD has become increasingly important to reduce the caliber of the venous cannula while maintaining an adequate blood flow [5–11]. We can affirm the VAVD has been a basic principle for the developing of minimally invasive surgical technique to correct heart disease.

Although the use of VAVD during MICS is used all over the world by adult and pediatric cardiac surgeons, its use is not uniform and varies according to the surgeon's or Institution experience.

Several reports in literature focused on studying the effect of the different negative pressures, applied to the system, on the flow produced by a cardiopulmonary bypass (CPB) circuit [8,11–15]. Kurusz and coworkers [15] tested different femoral cannula and assessed how their position (in inferior vena cava rather than in the right atrium) and the increasing negative pressure applied to the system affects the venous drainage flow. They concluded that the position of the cannula, rather than the design has effects on flow, that every cannula is capable of achieving higher flow when negative pressure is applied [8,11–15]. The combination of both siphon gravity and the applied vacuum pressure increase the drainage capabilities of each cannula.

Indications and the use of VAVD are not uniform and differ from center to center according to the Institutional experience. To standardize our clinical practice we attempted to determine the effects of suction on the venous drainage and consequently on the blood flow dynamic produced by the circuit. To do that we applied increasing vacuum suction to an experimental CPB circuit and measured the flow produced.

Our VAVD protocols bases on experimental flow evaluations showing the blood flow produced by each circuit according to the use of a specific cannula or a combination of cannulas and the degree of vacuum applied to the system (Tables 3.1–3.3). These tables are routinely used at the University of Padua in the pediatric cardiac surgery program to choose the best combination of venous cannulas and negative pressure according to the size of the surgical access and the size of the vena cava, to achieve the desired theoretical flow. However, we need to specify that these tables give indicative data since cannulas performance, for the same diameter, may vary depending of their design and flow-dynamics.

Fundamentals of Congenital Minimally Invasive Cardiac Surgery. http://dx.doi.org/10.1016/B978-0-12-811355-4.00003-4
Copyright © 2018 Elsevier Inc. All rights reserved.

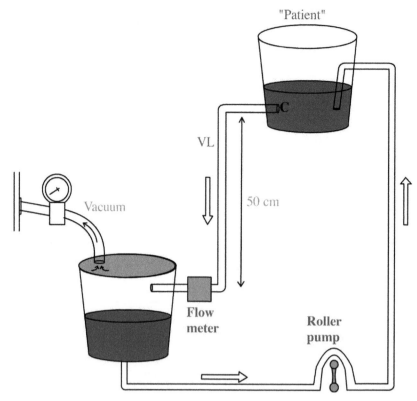

FIGURE 3.1 Schematic showing the cardiopulmonary bypass circuit routinely used during minimally invasive cardiac surgical procedure. *C*, Cannula; *VL*, venous line.

FIGURE 3.2 Wall vacuum system.

TABLE 3.1 Venous Drainage in mL/min According to Different Venous Cannulas and to the Type of Drainage (Syphon Gravity of Vacuum Assisted Drainage) Utilizing a 1/4″ Venous Line

| Cannulas | Syphon gravity (50 cm) | Vacuum assisted drainage (mmHg) | | | | | |
| | | −10 | −20 | −30 | −40 | −50 | −60 |
| --- | --- | --- | --- | --- | --- | --- | --- |
| 14 S | 670 | 790 | 880 | 980 | 1070 | 1160 | 1240 |
| 16 S | 720 | 830 | 1000 | 1150 | 1280 | 1400 | 1500 |
| 18 S | 800 | 1000 | 1150 | 1300 | 1450 | 1600 | 1750 |
| 20 S | 880 | 1100 | 1280 | 1460 | 1620 | 1770 | 1900 |
| 12 A-12 A | 740 | 920 | 1080 | 1180 | 1330 | 1430 | 1520 |
| 12 A-14 A | 820 | 1000 | 1160 | 1270 | 1430 | 1540 | 1630 |
| 14 A-14 A | 880 | 1060 | 1230 | 1360 | 1520 | 1650 | 1750 |
| 14 A-16 A | 950 | 1120 | 1300 | 1430 | 1600 | 1740 | 1850 |
| 16 A-16 A | 1000 | 1160 | 1340 | 1500 | 1670 | 1820 | 1950 |
| 16 A-18 A | 1100 | 1200 | 1390 | 1560 | 1730 | 1890 | 2030 |
| 18 A-18 A | 1150 | 1230 | 1420 | 1610 | 1780 | 1950 | 2100 |
| 18 A-20 A | 1200 | 1260 | 1480 | 1670 | 1850 | 2020 | 2190 |
| 20 A-20 A | 1250 | 1300 | 1520 | 1730 | 1910 | 2080 | 2250 |

Medtronic metal tipped angled cannulas (A), Medtronic straight cannulas (S) and the bicaval combinations of the two cannulas.

TABLE 3.2 Venous Drainage in mL/min According to Different Venous Cannulas and to the Type of Drainage (Syphon Gravity of Vacuum Assisted Drainage) Utilizing a 5/16″ Venous Line

| Cannulas | Syphon gravity (50 cm) | Vacuum assisted drainage (mmHg) | | | | | |
| | | −10 | −20 | −30 | −40 | −50 | −60 |
| --- | --- | --- | --- | --- | --- | --- | --- |
| 14 S | 800 | 1140 | 1280 | 1360 | 1440 | 1500 | 1580 |
| 16 S | 1000 | 1380 | 1560 | 1680 | 1800 | 1900 | 2020 |
| 18 S | 1280 | 1690 | 1900 | 2050 | 2200 | 2330 | 2470 |
| 20 S | 1570 | 2000 | 2230 | 2400 | 2580 | 2720 | 2880 |
| 12 A-12 A | 1400 | 1600 | 1790 | 1930 | 2060 | 2180 | 2300 |
| 12 A-14 A | 1500 | 1900 | 2090 | 2240 | 2380 | 2520 | 2650 |
| 14 A-14 A | 1600 | 2100 | 2290 | 2440 | 2590 | 2760 | 2900 |
| 14 A-16 A | 1800 | 2300 | 2520 | 2700 | 2880 | 3080 | 3250 |
| 16 A-16 A | 1900 | 2400 | 2640 | 2840 | 3040 | 3250 | 3450 |
| 16 A-18 A | 2000 | 2550 | 2820 | 3050 | 3280 | 3530 | 3750 |
| 18 A-18 A | 2200 | 2650 | 2910 | 3140 | 3350 | 3640 | 3850 |
| 18 A-20 A | 2400 | 2800 | 3080 | 3330 | 3600 | 3870 | 4100 |
| 20 A-20 A | 2600 | 3050 | 3300 | 3550 | 3800 | 4050 | 4250 |

Medtronic metal tipped angled cannulas (A), Medtronic straight cannulas (S) and the bicaval combinations of the two cannulas.

**TABLE 3.3** Venous Drainage in mL/min According to Different Venous Cannulas and to the Type of Drainage (Siphon Gravity of Vacuum Assisted Drainage) Utilizing a 3/8″ Venous Line

| Cannulas | Syphon gravity (50 cm) | Vacuum assisted drainage (mmHg) | | | | | |
| --- | --- | --- | --- | --- | --- | --- | --- |
| | | −10 | −20 | −30 | −40 | −50 | −60 |
| 16 S | 1200 | 1520 | 1700 | 1850 | 1950 | 2000 | 2050 |
| 18 S | 1560 | 2080 | 2370 | 2600 | 2700 | 2800 | 2860 |
| 20 S | 1820 | 2460 | 2800 | 3050 | 3250 | 3400 | 3500 |
| 16 A | 1400 | 1900 | 2200 | 2400 | 2500 | 2600 | 2650 |
| 18 A | 1750 | 2400 | 2750 | 3000 | 3150 | 3300 | 3400 |
| 20 A | 2050 | 2700 | 3050 | 3350 | 3550 | 3750 | 3900 |
| 24 A | 2800 | 3500 | 4000 | 4450 | 4850 | 5200 | 5500 |
| 16 A-16 A | 2400 | 2900 | 3300 | 3580 | 3850 | 4100 | 4300 |
| 16 A-18 A | 2600 | 3200 | 3600 | 4000 | 4300 | 4550 | 4800 |
| 18 A-18 A | 2800 | 3450 | 3900 | 4310 | 4700 | 4980 | 5160 |
| 18 A-20 A | 3000 | 3650 | 4100 | 4510 | 4900 | 5150 | 5380 |
| 20 A-20 A | 3200 | 3800 | 4250 | 4650 | 5000 | 5350 | 5580 |
| 20 A-24 A | 3500 | 4360 | 4890 | 5340 | 5750 | 6100 | 6450 |
| 24 A-24 A | 3800 | 4660 | 5230 | 5730 | 6220 | 6640 | 7000 |

Medtronic metal tipped angled cannulas (A), Medtronic straight cannulas (S) and the bicaval combinations of the two cannulas.

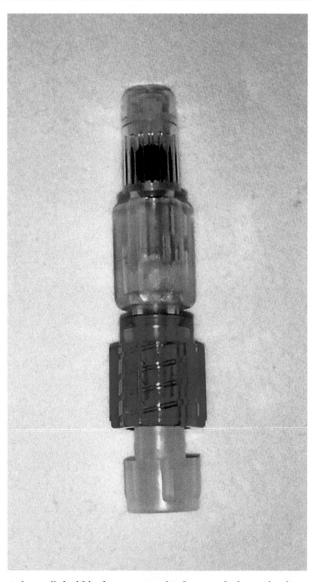

FIGURE 3.3 Additional safety valve to be applied within the vacuum assisted venous drainage circuit.

## SAFETY CONSIDERATIONS AND WARNINGS

Vacuum application is a widely used and practiced technique, but the related risks need not be underestimated. Safety and security systems must be used to avoid the pressurization of the venous reservoir and the production of excessive negative pressure which can damage the corpusculated blood part, generate gas microemboli and in extreme cases also damage the structure of the reservoir itself. Today's devices on the market are almost all equipped with premounted overpressure and antiimplosion valves, nonetheless, possibly additional valves are applicable (Fig. 3.3). In addition, a continuous monitoring with acoustic alert of the pressure measured directly in the venous reservoir (Fig. 3.4) is fundamental. In fact, under normal circumstances a slight discrepancy between the pressure measured by the vacuum system and the actual pressure controller

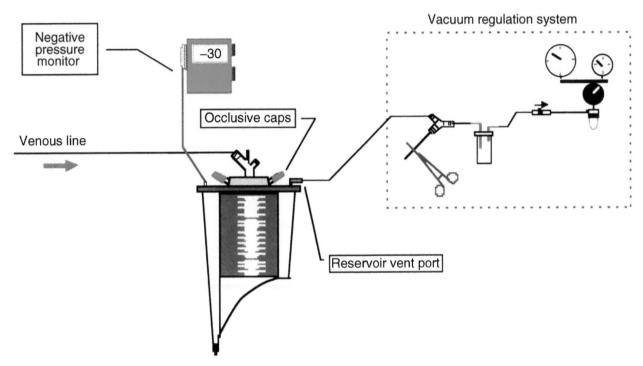

FIGURE 3.4    Circuit diagram for vacuum assisted venous drainage with direct reading in reservoir.

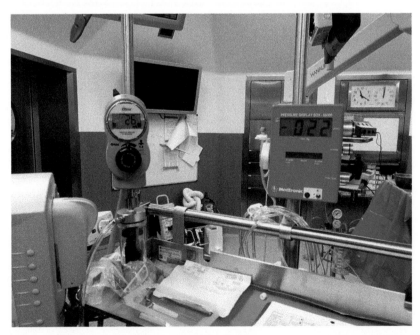

FIGURE 3.5    **Negative pressure regulation and monitoring system.** *Blue device*, vacuum system pressure measurement (vacuum regulator PUSH-TO-SET NEO 1331, Ohio Medical, Gurnee, IL, United States); *red device*, pressure measurement controller (pressure display box, Medtronic Inc, Minneapolis, MI, United States).

**FIGURE 3.6** **Triple pressure gauge monitoring system and expandable bag (Boehringer suction regulator model 7700, Boehringer, Germany).**

may be present (Fig. 3.5), but in the event of a controller malfunction a direct reading in the reservoir is necessary. It is also possible to connect an empty collapsible bag (Fig. 3.6), which in case of overpressure, swells, and therefore, in addition to allowing a partial discharge of the pressure, it acts as a visual indicator.

# REFERENCES

[1] Taketani S, Sawa Y, Masai T, et al. A novel technique for cardiopulmonary bypass using vacuum system for venous drainage with pressure relief valve: an experimental study. Artif Organs 1998;22:337–41.

[2] Durandy Y. Perfusionist strategies for blood conservation in pediatric cardiac surgery. World J Cardiol 2012;26(2):27–33.

[3] Sistino JJ, Michler RE, Mongero LB. Laboratory evaluation of a low prime closed-circuit cardiopulmonary bypass system. J Extra Corpor Technol 1993;24:116–9.

[4] Bevilacqua S, Matteucci S, Ferrarini M, et al. Biochemical evaluation of vacuum-assisted venous drainage: a randomized, prospective study. Perfusion 2002;17(1):57–61.

[5] Gregoretti S. Suction-induced hemolysis at various vacuum pressures: implications for intraoperative blood salvage. Transfusion 1996;36:57–60.

[6] Vida VL, Padalino MA, Motta R, Stellin G. Minimally invasive surgical options in pediatric heart surgery. Expert Rev Cardiovasc Ther 2011;9(6):763–9.

[7] Vida VL, Padalino MA, Bhattarai A, Stellin G. Right posterior-lateral mini-thoracotomy access for treating congenital heart disease. Ann Thorac Surg 2011;92(6):2278–80.

[8] Fiorucci A, Gerometta PS, DeVecchi M, Guzman C, Costantino ML, Arena V. In vitro assessment of the vacuum-assisted venous drainage (VAVD) system: risks and benefits. Perfusion 2004;19(2):113–7.

[9] Vida VL, Padalino MA, Boccuzzo G, et al. Minimally invasive surgery for congenital heart disease: a gender differentiated approach. J Thorac Cardiovasc Surg 2009;138(4):933–6.

[10] Nakanishi K, Shichijo T, Shinkawa Y, et al. Usefulness of vacuum-assisted cardiopulmonary bypass circuit for pediatric open-heart surgery in reducing homologous blood transfusion. Eur J Cardiothorac Surg 2001;20:233–8.

[11] Shin H, Yozu R, Maehara T, et al. Vacuum assisted cardiopulmonary bypass in minimally invasive cardiac surgery: its feasibility and effects on haemolysis. Artif Organs 2000;24:450–3.

[12] Lau CL, Posther KE, Stephenson GR, et al. Mini-circuit cardiopulmonary bypass with vacuum assisted venous drainage. Feasibility of an asanguineous prime in the neonate. Perfusion 1999;14:389–96.

[13] Kiyama H, Imazeki T, Katayama Y, Murai N, Mukouyama M, Yamauti N. Vacuum-assisted venous drainage in single-access minimally invasive cardiac surgery. J Artif Organs 2006;6(1):20–4.

[14] Colangelo N, Torracca L, Lapenna E, Moriggia S, Crescenzi G, Alfieri O. Vacuum-assisted venous drainage in extrathoracic cardiopulmonary bypass management during minimally invasive cardiac surgery. Perfusion 2006;21(6):361–5.

[15] Kurusz M, Deyo DJ, Sholar AD, Tao W, Zwischenberger JB. Laboratory testing of femoral venous cannulae: effect of size, position and negative pressure on flow. Perfusion 1999;14:379–87.

# FURTHER READING

[16] Munster K, Andersen U, Mikkelsen J, Pettersson G. Vacuum-assisted venous drainage (VAVD). Perfusion 1999;14:419–23.

# Chapter 4

# Surgical Instruments and Materials

Annalisa Francescato, Roberta Cabianca, Alvise Guariento

*University of Padua, Padua, Italy*

## SURGICAL INSTRUMENTS AND MATERIAL FOR MINIMALLY INVASIVE SURGICAL APPROACHES

See Images 4.1–4.25 and Video 4.1

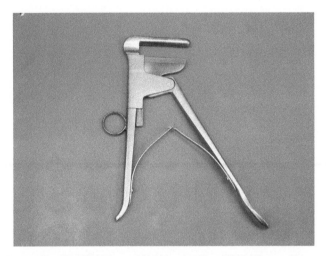

IMAGE 4.1    Schumacher's scissor for sternum AE FB911R mm 210 (Aesculap Inc., USA) (Chapter 8).

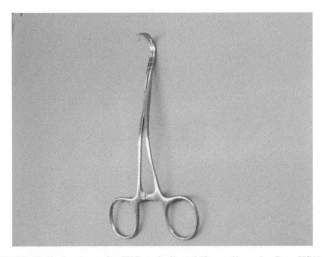

IMAGE 4.2    Satinsky clamp AE FB560R (Satinsky clamp for IVC encircling) 160 mm (Aesculap Inc., USA) (Chapters 8–10).

Fundamentals of Congenital Minimally Invasive Cardiac Surgery. http://dx.doi.org/10.1016/B978-0-12-811355-4.00004-6
Copyright © 2018 Elsevier Inc. All rights reserved.

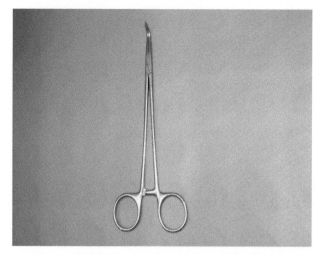

IMAGE 4.3    Hemostatic baby-Mixter forceps AE BJ012R (for SVC encircling)180–228 mm (Aesculap Inc., USA) (Chapters 8–10).

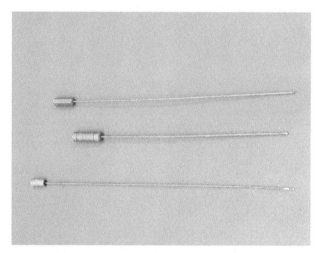

IMAGE 4.4    "Spike" adult suspension set (Iatrotek srl, Campaolongo Maggiore, VE, Italy) (Chapters 9–10).

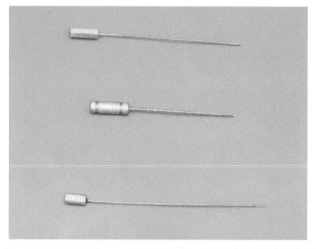

IMAGE 4.5    "Spike" infant suspension set (Iatrotek srl, Campaolongo Maggiore, VE, Italy) (Chapters 9–10).

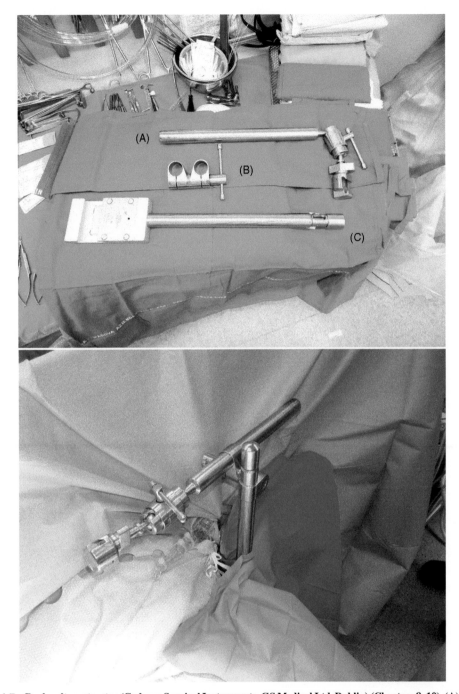

IMAGES 4.6 AND 4.7   **Bookwalter retractor (Codman Surgical Instruments, GS Medical Ltd, Dublin) (Chapters 8–10).** (A) Bookwalter horizontal bar; (B) Bookwalter junction; (C) Bookwalter table post.

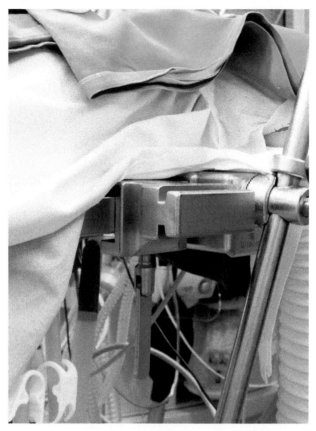

IMAGE 4.8    Bookwalter table support—COD 8060200 (Codman Surgical Instruments, GS Medical Ltd, Dublin) (Chapters 8–10).

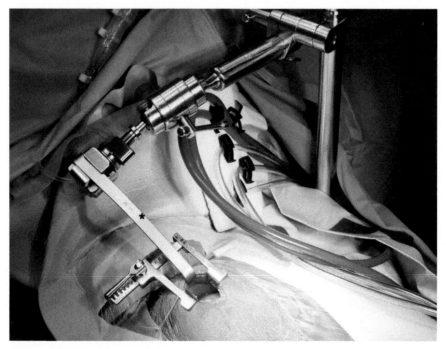

IMAGE 4.9    Farabeuf retractor (*) (105–170 mm) (Croma Gio.Batta Surgical Instruments, Padova, Italy) and Pilling coastal autostatic divaricator (Finocchietto retractor) from 140 × 160 to 170 × 200 mm (Teleflex Medical Europe Ltd, Ireland) (Chapters 8–10).

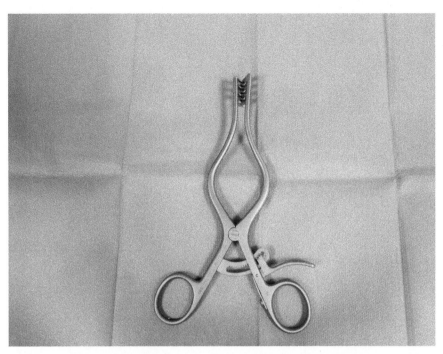

IMAGE 4.10　Weilaner autostatic retractor for femoral vessels isolation—AE BV 070R (small, 105 mm) or BV 075R (medium, 165 mm) (Aesculap Inc., USA) (Chapter 7).

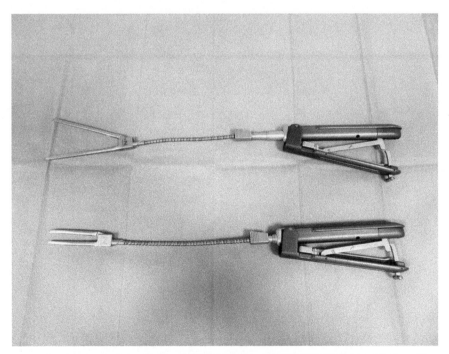

IMAGE 4.11　Novare Cygnet Straight Aortic Cross Clamp—H0132 N10183 (small)/H0184 N10142 (big) (Vitalitec, Plymouth, MA, USA).

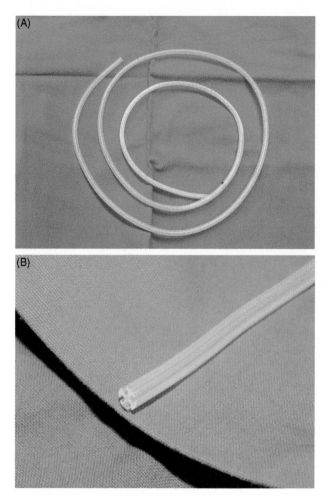

IMAGE 4.12  (A) Silicone Vac drain, round (15-19-24 mm) (Fortune Medical Instruments Corp., New Taipei City, Taiwan). (B) Detail of the tip of the chest tube showing its cross-section.

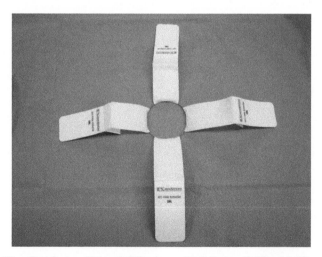

IMAGE 4.13  **Soft tissue retractor, Thru Port System (Edwards Lifesciences LLC, Irvine, California) (Chapters 9–10).**

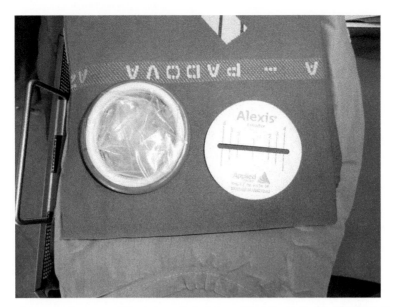

IMAGE 4.14  Alexis wound protector/retractor (XS: 2–4 cm; S: 2, 5–6 cm; M: 5–9 cm) (Applied Medical, Rancho Santa Margherita, USA) (Chapters 9–10).

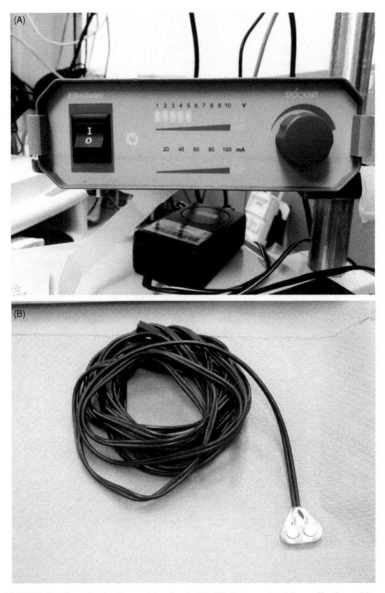

IMAGE 4.15  (A) Fibrillator Fi 20 M, Stockert (Sorin Group, Munchen); (B) Fibrillator potential equalization cable and electrode, Stockert (Sorin, Munchen) (Chapters 9–10).

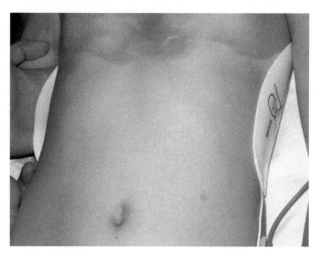

IMAGE 4.16    Heartstart Pads positioned on each side of the patients (Infant <10 kg, Adult >10 kg) (Philips Medical System, Andover, MA, USA) (Chapters 8–10).

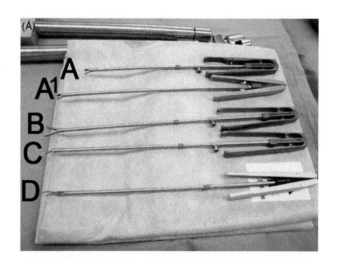

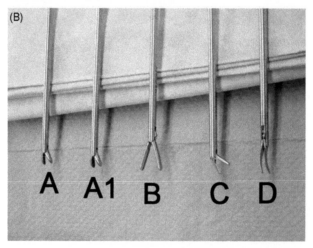

IMAGE 4.17    (A and B) Port-access minimally invasive surgical kit. A Castroviejo Needle Holder (*short*) (CardioVations, Edwards Lifesciences LLC, Irvine, California); A1 Castroviejo Needle Holder (*long*) (CardioVations, Edwards Lifesciences LLC, Irvine, California); B Forcep (delicate tip) (CardioVations, Edwards Lifesciences LLC, Irvine, California); C Curved surgical forceps (CardioVations, Edwards Lifesciences LLC, Irvine, California) (CardioVations, Edwards Lifesciences LLC, Irvine, California); D Straight surgical forceps.

IMAGE 4.18    Pediatric arterial cannula and introducers (Medtronic Bio-Medicus, Medtronic Inc., Minneapolis, MN, USA).

IMAGE 4.19    Pediatric venous cannula and introducers (Medtronic Bio-Medicus, Medtronic Inc., Minneapolis, MN, USA).

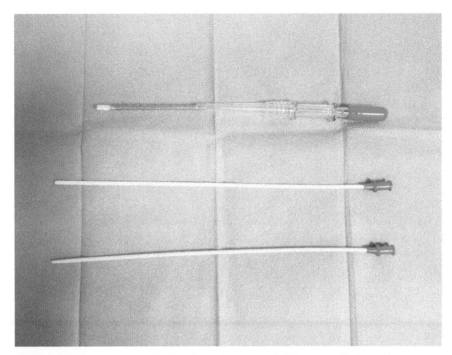

IMAGE 4.20    Fem-Flex II femoral arterial cannula with Duraflo coating (Edwards Lifesciences, Irvine, CA).

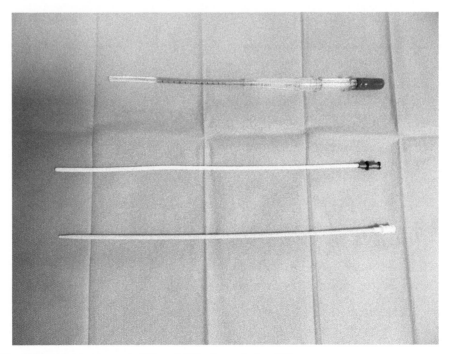

IMAGE 4.21    Fem-Flex II femoral venous cannula (Edwards Lifesciences, Irvine, CA).

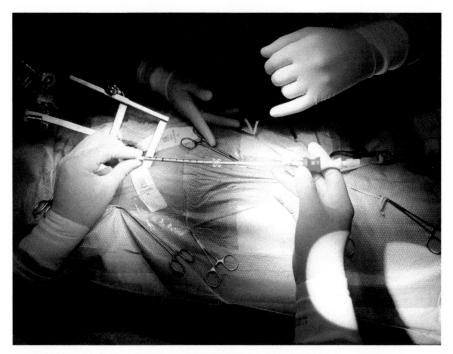

IMAGE 4.22 **Arterial cannulas—bio-medicus pediatric arterial cannulas and introducer (Medtronic, Minneapolis, USA) (8–14 French) and Optisite arterial perfusion cannulas (Edwards Lifescences, Inc., Irvine USA) (8–22 French)**.

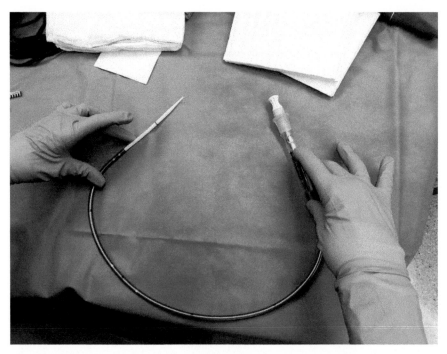

IMAGE 4.23 **Femoral venous cannulas:** Bio-medicus Pediatric femoral venous cannula and introducer (Medtronic, Minneapolis, USA) (8–14 French), Bio-medicus Adult femoral venous cannula and introducer (Medtronic, Minneapolis, USA) (15–25 French). Internal Jugular vein cannulas: Bio-medicus adult cannula and introducer (Medtronic, Minneapolis, USA) (15–23 French).

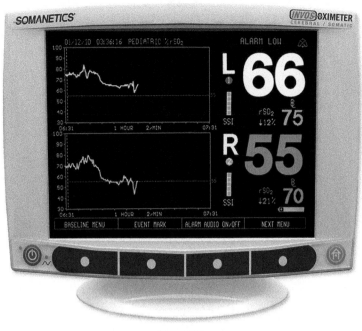

2-Channel Monitor

IMAGE 4.24 **The INVOS 5100C oximeter (Covidien, Mansfield, MA, US)**.

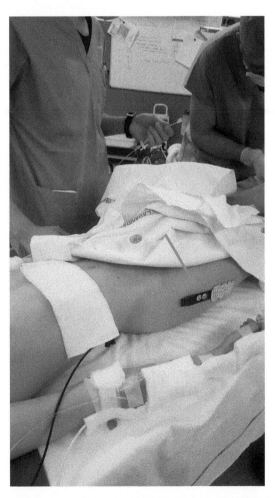

IMAGE 4.25 **INVOS sensors**. Neonatal sensor for patients with body weight ≤5 kg (OxyAlertTM NIRS-Cerebral neonatal regional oxygen saturation sensor, COVIDIEN, Mansfield, MA, US) Pediatric sensor for patients with body weight >5 kg (SomaSensor, Pediatric Cerebral/Somatic sensor, COVIDIEN, Mansfield, MA, US).

VIDEO 4.1 The "SPIKE" pericardial suspension set.

## FURTHER READING

[1] Vida VL, Padalino MA, Boccuzzo G, Veshti AA, Speggiorin S, Falasco G, et al. Minimally invasive operation for congenital heart disease: a sex-differentiated approach. J Thorac Cardiovasc Surg 2009;138:933–6.

[2] Vida VL, Padalino MA, Motta R, Stellin G. Minimally invasive surgical options in pediatric heart surgery. Expert Rev Cardiovasc Ther 2011;9:763–9.

[3] Vida VL, Tessari C, Fabozzo A, Padalino MA, Barzon E, Zucchetta F, et al. The evolution of the right anterolateral thoracotomy technique for correction of atrial septal defects: cosmetic and functional results in prepubescent patients. Ann Thorac Surg 2013;95:242–7.

Chapter 5

# Operative Echocardiography

Demetrio Pittarello, Karmi Shafer
*University of Padua, Padua, Italy*

## INTRODUCTION

Transesophageal echocardiography (TEE) has a major role in the optimal surgical management of pediatric cardiac surgery as it gives useful information during all procedural phases [1] (Tables 5.1–5.2).

1. Preoperatively (before cardiopulmonary bypass—CPB): it is used to delineate cardiac anatomy and structural details as confirmation of the diagnoses previously established by transthoracic echocardiography (TTE) and angiography. It is particularly suited to define complex anatomical structures, functional abnormalities, and flow disturbances that may not always be obtainable from TTE alone [2,3], thus identifying possible additional pathologic conditions;
2. Intraoperatively (during CPB): it is useful to guide the deairing maneuvers to show cardiac function during CPB weaning and optimize fluid challenge;
3. Postoperatively (after weaning of CPB): it is useful to evaluate surgical repair, assess residual lesions, determine the need for reintervention or surgical revision, evaluate myocardial performance, estimate cardiac function, and determine the need for inotropic treatment [4–12].

## EPICARDIAL ECHOCARDIOGRAPHY

Epicardial echocardiography was one of the first techniques used to evaluate the immediate postoperative results, and, indeed, provided some valuable information [13]. This technique has the advantage of utilizing the whole armamentarium of transducers; however, it requires the surgeon to operate the transducer, which is placed directly on the heart, thereby having the potential for arrhythmias, hemodynamic compromise, and infection. Inadequate ultrasound windows also can be a problem in the neonate, as there is a very limited space to apply the probe (Table 5.3), particularly during mini-invasive surgery.

Currently, epicardial echocardiography has virtually been replaced by TEE that allows the serial assessment of anatomy and function without interruption of surgery. TEE is performed prebypass to confirm or modify preoperative findings and assist in anesthetic and surgical planning [5,14].

Initially, many surgeons were skeptical about the use of TEE; however, with time, they have come to realize that the benefits of this technique far outweigh any small risks to the patient, including the additional time needed to perform the studies. TEE now is the standard of practice in the management of the adult and child undergoing cardiac surgical intervention.

However, despite the limitations of epicardial echocardiography, it still has a role in those cases where TEE cannot be utilized, such as in cases with esophageal abnormalities or the occasional neonate in whom the pediatric probe cannot be inserted.

## PROBES AND VIEWS

The TEE probe is inserted before the beginning of the operation by an anesthetist or a cardiologist. Generally, the probe is inserted after induction of general anesthesia using the following parameters: in patients with a weight >20 kg, an adult probe (xMatrix probe: X7-2t; 3D matrix array probe; 2-7 MHx; Philips, Andover, MA) with 2500 elements per transducer; in pediatric patients, who weigh >3.5 kg, a pediatric probe (Mini Multi probe: S7-3t; 3–7 MHz; Philips, Andover, MA)

**TABLE 5.1 Intraoperative Pediatric Echocardiography**

- TEE is performed *prebypass* to confirm or modify preoperative findings and assist in anesthetic and surgical planning
- TEE is used during rewarming *on bypass* to initially asses surgical repair, ventricular function, impact of inotropic intervention, and to detect cavitary and myocardial air
- TEE is performed *postbypass* to asses surgical repair, for qualitative assessment of contractility and volume status

**TABLE 5.2 Intraoperative Evaluation With TEE of Atrial Septal Defect**

**Prebypass evaluation**

- Size
- Position and shunt direction
- RA and RV sizes

- Associated anomalies
- Mitral and tricuspid valve competence
- LV function

**Postbypass evaluation**

- Residual shunting
- Valve competence
- Ventricular preload

- Pulmonary and systemic veins obstruction
- LV function
- Air emboli

**TABLE 5.3 Epicardial Echocardiography (EE) Versus Transesophageal Echocardiography (TEE)**

| PROS | CONS |
|---|---|
| • Available technology in a novel manner<br>• Potentially better imaging windows (operator dependent) | • Arrhythmias<br>• Hemodynamic compromise<br>• Theoretical risk of infection<br>• EE requires the surgeon to stop working |

with 64 elements per transducer and dimensions of 10.7 mm (tip width) and 8.0 mm (tip height) (Fig. 5.1); in neonates, weighing from 2.0 kg to 3.5 kg, an infant biplane probe (Micro TEE probe: S8-3t; 3–8 MHz; Philips, Andover, MA) with 34 elements per transducer and dimensions of 7.5 mm (tip width) and 5.5 mm (tip height) (Fig. 5.2). In our experience, TEE examinations were performed using Philips Sonos IE33 echocardiography machines (Philips, Andover, MA) equipped with pulsed, continuous wave, and color Doppler capabilities, with the possibility to use the three-dimensional (3D) echocardiography, having the adult probe, at the anesthesiologist's discretion.

Standard views are obtained according to the guidelines of the American Society of Echocardiography (ASE) and the Society of Cardiovascular Anesthesiologists (SCA) [15], preoperatively and before the beginning of CPB and postoperative at the end of CPB before revising heparin and the closure of the chest (Fig. 5.3).

## SAFETY

When conducted properly, TEE is considered a safe procedure in the pediatric patient, and it has been documented in patients with less weight [16]. The miniaturization of transesophageal probes has allowed their easy placement in infants and children [11,13], and previous studies have demonstrated not only its safety [11,17,18], but also its utility [19,20] and cost-effectiveness [6] for routine use. To provide the best care for children undergoing open cardiac surgery, perioperative TEE is essential, and its routine use in pediatric cardiac surgery should no longer be a matter of choice.

In our experience, keeping the probe during the procedure until postoperative study is completed is considered a safe practice, whereas others advocate reinserting the probe at the end of the surgery [21].

As far as the weight of the pediatric patients is concerned, although adult TEE probes have been successfully used in patients weighing as less as 14.7 kg [22], some authors recommend using a pediatric TEE probe for all patients weighing <20 kg [5,22].

Except for patients weighing <2 kg and select patients with tracheal and esophageal anomalies [23], routine intraoperative transesophageal echocardiography can be used.

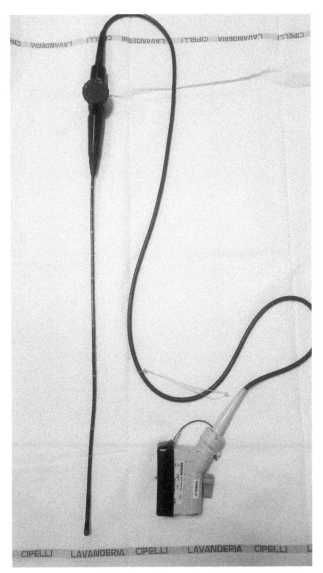

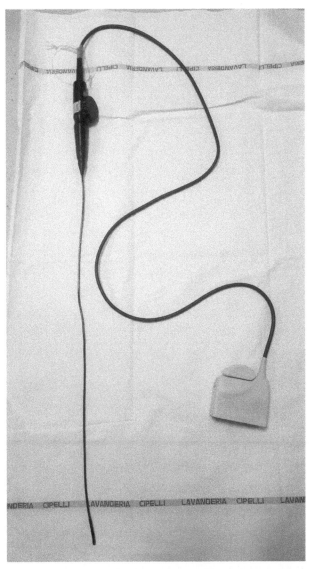

FIGURE 5.1 **Mini-multi TEE probe: S7-3t; 3–7 MHz; Philips, Andover, MA.** 64 elements per transducer and dimensions of 10.7 mm (tip width) and 8.0 mm (tip height).

FIGURE 5.2 **Micro TEE probe: S8-3t; 3–8 MHz; Philips, Andover, MA.** 34 elements per transducer and dimensions of 7.5 mm (tip width) and 5.5 mm (tip height).

In some cases, patients weighing as less as 2 kg have successfully and safely undergone intraoperative TEE [24]; however, additional caution is to be exercised when inserting a probe in a neonate weighing <3 kg. Although complications are rare, those most frequently encountered relate to oropharyngeal and esophageal traumas, including hoarseness and dysphagia after the procedure.

Several large series have reported a 1%–3% incidence of complications while performing TEE [6,25] (Table 5.4). Complications that are more common include failure to insert TEE probe or airway obstruction.

However, there are other important rare complications such as right main stem advancement of the endotracheal tube (0.2%), tracheal extubation (0.5%), and vascular compression (0.6%) [25].

The previously mentioned studies demonstrate that TEE can be safely performed in pediatric patients with minimal or no mucosal injury. Probe-related injuries include thermal pressure trauma, mechanical problems resulting in laceration, and/or perforation of the oropharynx, hypopharynx, esophagus, and stomach [26–29]. This may vary from minor trauma to catastrophic perforation of the esophagus [17,27,30], which, although extremely rare, is found in the literature in anecdotal case reports, with one case report in the pediatric age group [31].

Complications can be minimized by utilizing experienced individuals to intubate the esophagus in conjunction with using the correct-sized probe for the pediatric patient. Complications can also include arrhythmias, pulmonary complications

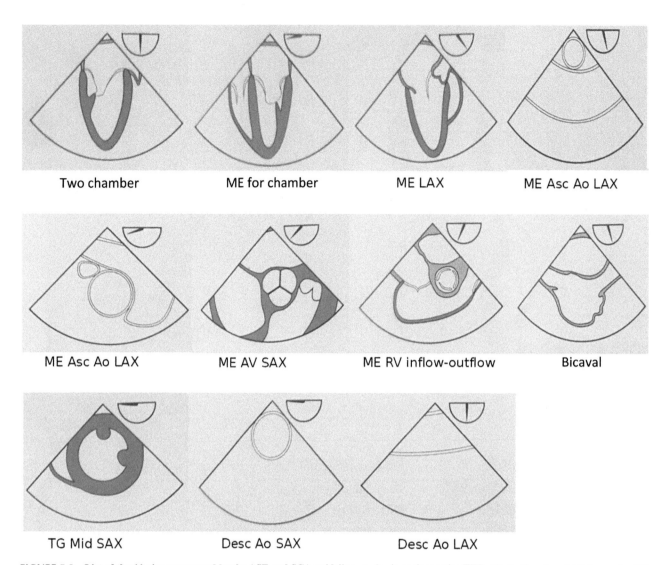

**FIGURE 5.3** **List of the 11 views suggested by the ASE and SCA guidelines on basic perioperative TEE.** *AV*, Aortic valve; *LAX*, long axis; *ME*, mid-esophageal; *RV*, right ventricle; *SAX*, short axis.

**TABLE 5.4 Patient Safety**

| 1%–3% Incidence of complications in pediatric population | |
| --- | --- |
| All complications | 3.2% |
| Failure to insert TEE probe | 0.8% |
| Airway obstruction | 0.8% |
| Advancement of endotracheal tube | 0.2% |
| Inadvertent tracheal extubation | 0.5% |
| Vascular compression | 0.6% |
| Other (gastric incision, etc) | 0.4% |

**Source:** Stevenson JG. Incidence of complications in pediatric transesophageal echocardiography: experience in 1650 cases. J Am Soc Echocardiogr 1999;12:527–32.

(bronchospasm, hypoxemia, laryngospasm), and circulatory derangement with hypotension. Between airway obstruction with increased peak inspiratory pressures [22,32] they are noted especially when we use the transgastric views, likely because the anteroflexion of the probe against the diaphragm. To avoid airway compression, in small infants who display compromised ventilation, positioning the probe in the hypopharynx while not actively imaging has been recommended [33]. For the 5% [34] in whom transesophageal echocardiographic imaging was not considered safe, because of low weight, airway compromise with probe insertion, or inherent tracheal and esophageal abnormalities, assessment was unfortunately limited.

However, in general, if an appropriate-sized probe is carefully inserted and manipulated, TEE in pediatric patients appears to be relatively safe.

## INDICATIONS

According to the practice guidelines for perioperative TEE established by the SCA and the American Society of Anesthesiologists (ASA) [35], there is strong evidence for the usefulness of TEE in surgery for congenital heart disease (CHD), and it has become a standard imaging technology for patients with CHD undergoing intervention in the operating room [6,10,19,21,36–39]. This procedure significantly improves the clinical outcome of these patients. Although it is now a common practice for anesthesiologists to be primarily responsible for the interpretation of intraoperative TEE during adult cardiac surgery, their role in TEE during congenital cardiac surgery is still being debated [40,41].

Usually, TEE is used in patients with CHD for assessment during cardiac surgery (Table 5.5).

The intraoperative TEE performed just before surgical intervention can provide additional information and may be of benefit [42]. It may confirm or exclude preoperative TTE findings and assess the immediate preoperative hemodynamics and ventricular function of the patient. In addition, the findings can be directly demonstrated to the surgeon and anesthesiologist for immediate review just prior to the commencement of the operation. Preoperative TEE may also facilitate the placement of central venous catheters, selection of anesthetic agents, and use of preoperative inotropic support by demonstrating ventricular systolic function and size [43,44].

## CONTRAINDICATIONS

Table 5.6 lists the absolute and relative contraindications to perform a TEE examination in the patient with CHD. The risk of the TEE procedure and benefits must be carefully weighed in patients with cervical and thoracic spinal abnormalities that may distort the normal orientation of the esophagus. Patients with Down syndrome, for example, have intrinsic narrowing

---

**TABLE 5.5 Current Indications for TEE in Patients With CHD**

1. Diagnostic indications
   - Patient with suspected CHD and nondiagnostic TTE
   - Presence of PFO and direction of shunting as possible etiology for stroke
   - PFO evaluation with agitated saline contrast to determine possible right-to-left shunt, prior to transvenous pacemaker insertion
   - Evaluation of intra- or extracardiac baffles following the Fontan, Senning, or Mustard procedure
   - Aortic dissection (Marfan syndrome)
   - Intracardiac evaluation for vegetation or suspected abscess
   - Pericardial effusion or cardiac function evaluation and monitoring postoperative patient with open sternum or poor acoustic windows
   - Evaluation for intracardiac thrombus prior to cardioversion for atrial flutter/fibrillation
   - Evaluation status of prosthetic valve
2. Perioperative indications
   - Immediate preoperative definition of cardiac anatomy and function
   - Postoperative surgical results and function
3. TEE-guided interventions
   - Guidance for placement of ASD or VSD occlusion device
   - Guidance for blade or balloon atrial septostomy
   - Catheter tip placement for valve perforation and dilation in catheterization laboratory
   - Guidance during radiofrequency ablation procedure
   - Results of minimally invasive surgical incision or video-assisted cardiac procedure

*ASD*, Atrial septal defect; *CHD*, congenital heart disease; *PFO*, patent foramen ovale; *TTE*, transthoracic echocardiography; *VSD*, ventricular septal defect.

**TABLE 5.6 Contraindications for Transesophageal Echocardiography**

| Absolute contraindications | Relative contraindications |
|---|---|
| • Unrepaired tracheoesophageal fistula<br>• Esophageal obstruction or stricture<br>• Perforated hollow viscus<br>• Poor airway control<br>• Severe respiratory depression<br>• Uncooperative, unsedated patient | • History of prior esophageal surgery<br>• Esophageal varices or diverticulum<br>• Gastric or esophageal Bleeding<br>• Vascular ring and aortic arch anomaly with or without airway compromise<br>• Oropharyngeal pathology<br>• Severe coagulopathy<br>• Cervical spine injury or anomaly |

**TABLE 5.7 TEE Influences in Intraoperative Surgical Management**

• Preoperative diagnosis changed in 3%–5% of the patients
• Overall surgical management changed due to intraoperative TEE in 7% of the patients (does not differ significantly from that reported with EE)

of the hypopharyngeal region, in addition to increased incidence of cervical spine anomalies, thereby resulting in difficult, or failed, probe insertion. In patients with previous esophageal surgery, history of dysphagia or significant coagulopathy, although associated with higher risk, is considered a relative contraindication [45].

## PRE-CPB UTILITY OF TEE

Optimal surgical repair requires accurate preprocedural evaluation followed by critical evaluation of the surgical repair by intraoperative TEE.

Preoperative TEE allows the confirmation of diagnoses established by TTE and angiography, as well as identifies possible additional pathologic conditions (Table 5.7).

Some centers, in addition to us, recommend the usage of preoperative TEE for all pump and nonpump cases [46]. Smallhorn suggested that complex repair, valve surgery, and outflow tract obstruction most benefited from preoperative evaluation, excluding simple lesions such as ASD, single VSD, and extracardiac defects [21].

Otherwise, in our experience, complete TEE examinations are conducted in all patients, as well as in those undergoing mini-invasive surgery and before CPB, to confirm or modify the interpretation of preoperative examinations (Figs. 5.4–5.5). In case of relevant modifications of the preoperative diagnosis, the pre-CPB TEE findings are displayed and discussed with the surgical team.

Other indications are to confirm placement of venous cannulae during mini-invasive approach and left ventricular venting device eliminating the possibility of left ventricular dilation because of inadequate venting.

At the end of the examination, before the beginning of CPB, the probe is advanced into the stomach and left in an unlocked position; during bypass, the ultrasound emission is turned off.

With regard to the planes, as we said, standardized imaging have been established for a comprehensive adult TEE examination [15]. The ASE/SCA has established guidelines and standards for performing a comprehensive TEE in the adult with heart disease, as well as basic guidelines to do a comprehensive TEE exam [47] (Fig. 5.3).

These views and planes may be applied to the pediatric patient as well, even if there is currently no uniform consensus on standardized views and planes to be used in the pediatric patient.

## POST-CPB UTILITY OF TEE

Although intraoperative TEE is helpful in reviewing the preoperative diagnosis of patients entering the OR and planning additional interventions, postoperative TEE evaluation assesses the global success of surgery.

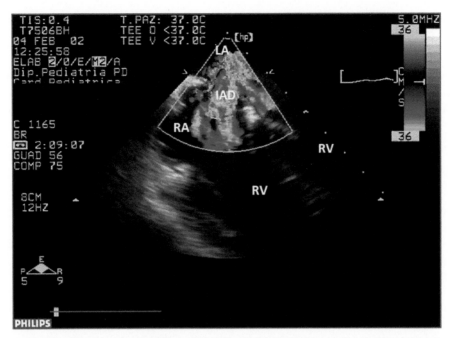

**FIGURE 5.4  Mid-esophageal four-chamber view of a patient with atrial septal defect.** All four cardiac chambers can be visualized in this view. Applications of this cross-section include the evaluation of chamber size and morphology, atrioventricular valve function, and intracardiac shunting across the atrial septum. Color Doppler interrogation of the interatrial septum in this view complements the evaluation of atrial-level communications. *IAD*, Interatrial defect; *LA*, left atrium; *LV*, left ventricle; *RA*, right atrium; *RV*, right ventricle.

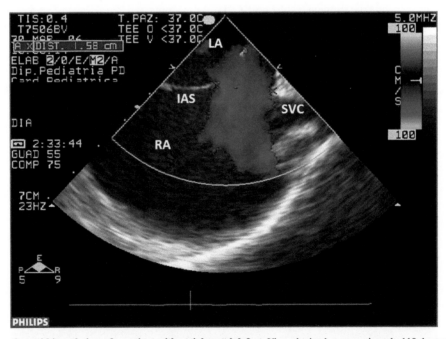

**FIGURE 5.5  Mid-esophageal bicaval view of a patient with atrial septal defect.** View obtained at approximately 110 degree in the mid-esophagus displays the right atrium, a portion of the left atrium, the interatrial septum, and the superior vena cava. Color Doppler interrogation of the interatrial septum in this view complements the evaluation of atrial-level communications. This view is also useful in optimizing catheter position during central venous line placement. *IAS*, Interatrial septum; *LA*, left atrium; *RA*, right atrium; *SVC*, superior vena cava.

There are numerous phases in which the TEE is used, with some as follows:

- Before declamping the aorta and during rewarming on CPB, it is possible to detect cavitary air and lead the surgeon to an appropriate deairing.
- Before disconnection of bypass cannulae, it provides the opportunity to detect significant and potentially treatable disease, determine the need for reintervention, and evaluate myocardial performance and contractility, volume status, and need for inotropic treatment. The need to return to bypass after initial surgical intervention is a serious decision that is influenced not only by the intraoperative transesophageal imaging but also by direct visualization and assessment by the surgeon.
- Before transfer to the intensive care unit, it allows the evaluation of cardiac sufferance during sternal closure, the presence of pericardial effusion, and any new changes of cardiac performance, thereby preventing reoperation at a later date, further morbidity and mortality, or increased hospital length of stay and higher costs.

The transgastric short- and long-axis views (Fig. 5.6) of the left and right ventricles are the optimal positions for evaluation of myocardial function, as they permit an assessment of the majority of the myocardial segments. Although usually ejection fraction is not calculated, it is relatively easy to provide a qualitative global assessment of function. Likewise, segmental wall motion abnormalities can be appreciated.

Although the majority of pediatric patients are not undergoing coronary artery surgery, segmental abnormalities can occur due to intracoronary air or localized reperfusion injury. A qualitative assessment of filling also is possible and invariably is helpful to the anesthetist and perfusionist attempting to optimize preload.

When a residual lesion is identified, it is crucial that the imaging is sufficiently clear to lead constructive discussion between surgeon and anesthesiologists as to whether to recommence CPB [14]. For this reason, it is essential for the anesthesiologist to have significant involvement in TEE. In our institution, the perioperative team consists of the surgeon, anesthesiologist, pediatric cardiologist performing the initial examination and postrepair examination, and an additional expert help, when needed.

Problems can occur while a patient is still on CPB, and an informed anesthesiologist can alert the surgeon to problems that can be handled and then request additional help from the cardiologist. The ability to repair the residual lesion should be weighed against the consequences of further period of CPB.

Furthermore, in patients with minimally invasive cardiac surgical procedures where direct visualization of the heart is limited, TEE assists in evaluating the surgical results post-CPB. For example, the assessment of an ASD closure is relatively straightforward and involves the visualization of the atrial patch to ensure that there are no significant residual shunts. Evaluation of the tricuspid valve and a qualitative assessment of biventricular function are also advisable.

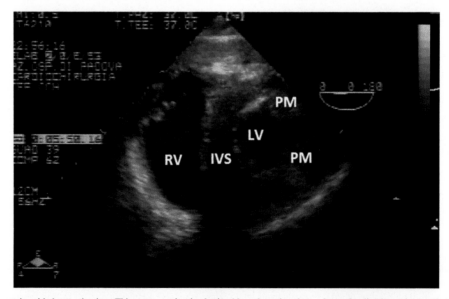

FIGURE 5.6 Transgastric mid short-axis view. This cross-section is obtained by advancing the probe to the distal esophagus. Anteflexion is required to maintain adequate probe contact. This view provides a quick assessment of ventricular filling, as well as global and segmental function. This mid-papillary ventricular cross-section is particularly useful in determining the etiology of hypotension upon separation from cardiopulmonary bypass. *IVS*, Interventricular septum; *LV*, left ventricle; *PM*, papillary muscle; *RV*, right ventricle.

## 3D VISUALIZATION

Three-dimensional (3D) TEE is extensively used nowadays in adult cardiac surgery, especially for the disease of the mitral valve [45,47,48]. Although it has been described in pediatric patients weighing >20 kg (where we can use the adult probe), a miniaturized 3D TEE probe is not yet available for the pediatric population, and it would be necessary to produce the probe by factory. There are promising elements for the future, but the use of more advanced imaging modalities such as three-dimensional echocardiography in the pediatric setting is not fully available at the moment.

## CONCLUSIONS

In conclusion, routine TEE imaging is helpful in the surgical repair of CHD in children and adults. Performance of TEE in the patient with CHD submitted to mini-invasive surgery immediately before and after surgery and before chest closure has been a contributor to the overall excellence in outcome for congenital heart surgery achieved in the past decade. Based on the TEE and clinical findings, the surgeon, in conjunction with the TEE anesthesiologist–echocardiographer, determines whether the surgical repair is acceptable or needs acute revision.

These clinical benefits, a growing number of highly skilled operators, and improvements in technology have lead to rapid adoption of TEE monitoring in pediatric cardiac surgery with remarkable advances in the management of patients with CHD.

## REFERENCES

[1] Sheil ML, Baines DB. Intraoperative transoesophageal echocardiography for paediatric cardiac surgery-an audit of 200 cases. Anaesth Intensive Care 1999;27:591–5.

[2] Kamra K, Russell I, Miller-Hance WC. Role of transesophageal echocardiography in the management of pediatric patients with congenital heart disease. Pediatr Anesth 2011;21:479–93.

[3] Jijeh AMZ, Omran AS, Najm HK, Abu-Sulaiman RM. Role of intraoperative transesophageal echocardiography in pediatric cardiac surgery. J Saudi Heart Assoc 2016;28:89–94.

[4] Bettex DA, Schmidlin D, Bernath MA, et al. Intraoperative transesophageal echocardiography in pediatric congenital cardiac surgery: a two-center observational study. Anesth Analg 2003;97:1275–82.

[5] Muhiudeen IA, Roberson DA, Silverman NH, Haas GS, Turley K, Cahalan MK. Intraoperative echocardiography for evaluation of congenital heart defects in infants and children. Anesthesiology 1992;76:165–72.

[6] Randolph GR, Hagler DJ, Connolly HM, et al. Intraoperative transesophageal echocardiography during surgery for congenital heart defects. J Thorac Cardiovasc Surg 2002;124:1176–82.

[7] Mochizuki Y, Patel AK, Banerjee A, et al. Intraoperative transesophageal echocardiography: correlation of echocardiographic findings and surgical pathology. Cardiol Rev 1999;7:270–6.

[8] Kaushal SK, Dagar KS, Singh A, et al. Intraoperative echocardiography as a routine adjunct in assessing repair of congenital heart defects: experience with 300 cases. Ann Card Anaesth 1998;1:36–45.

[9] Loick HM, Scheld HH, Van Aken H. Impact of perioperative transesophageal echocardiography on cardiac surgery. Thorac Cardiovasc Surg 1997;45:321–5.

[10] O'Leary PW, Hagler DJ, Seward JB, et al. Biplane intraoperative transesophageal echocardiography in congenital heart disease. Mayo Clin Proc 1995;70:317–26.

[11] Stevenson JG, Sorensen GK, Gartman DM, et al. Transesophageal echocardiography during repair of congenital cardiac defects: identification of residual problems necessitating reoperation. J Am Soc Echocardiogr 1993;6:356–65.

[12] Benheim A, Karr SS, Sell JE, et al. Routine use of transesophageal echocardiography and color flow imaging in the evaluation and treatment of children with congenital heart disease. Echocardiography 1993;10:583–93.

[13] Muhiudeen IA, Roberson DA, Silverman NH, et al:. Intraoperative echocardiography in infants and children with congenital cardiac shunt lesions: transesophageal versus epicardial echocardiography. J Am Coll Cardiol 1990;16:1687–95.

[14] Stevenson JG. Role of intraoperative transesophageal echocardiography during repair of congenital cardiac defects. Acta Paediatr Suppl 1995;410:23–33.

[15] Hahn RT, Abraham T, Adams MS, Bruce CJ, Glas KE, Lang RM, Reeves ST, Shanewise JS, Siu SC, Stewart W, Picard MC. Guidelines for performing a comprehensive transesophageal echocardiographic examination: recommendations from the American Society of Echocardiography and the Society of Cardiovascular Anesthesiologists. J Am Soc Echocardiogr 2013;26:921–64.

[16] Hilberath JN, Oakes DA, Shernan K, Bulwer BE, D'Ambra MN, Eltzschig HK. Safety of transesophageal echocardiography. J Am Soc Echocardiogr 2010;23:1115–27.

[17] Daniel WG, Erbel R, Kasper W, et al. Safety of transesophageal echocardiography. A multicenter survey of 10,419 examinations. Circulation 1991;83:817–21.

[18] Hilberath JN, Oakes DA, Shernan SK, Bulwer BE, D'Ambra MN, Eltzschig HK. Safety of transesophageal echocardiography. J Am Soc Echocardiogr 2010;23:1115–27. quiz 220-1.

[19] Bengur AR, Li JS, Herlong JR, Jaggers J, Sanders SP, Ungerleider RM. Intraoperative transesophageal echocardiography in congenital heart disease. Semin Thorac Cardiovasc Surg 1998;10:255–64.

[20] Peng DM, Sun HY, Hanley FL, Olson I, Punn R. Coronary sinus obstruction after atrioventricular canal defect repair. Congenit Heart Dis 2014;9:E121–4.

[21] Smallhorn JF. Intraoperative transesophageal echocardiography in congenital heart disease. Echocardiography 2002;19:709–23.

[22] Stevenson JG, Sorensen GK. Proper probe size for pediatric transesophageal echocardiography. Am J Cardiol 1993;72:491–2.

[23] Ayres NA, Miller-Hance W, Fyfe DA, et al. Indications and guidelines for performance of transesophageal echocardiography in the patient with pediatric acquired or congenital heart disease: report from the task force of the Pediatric Council of the American Society of Echocardiography. J Am Soc Echocardiogr 2005;18:91–8.

[24] Mart CR, Fehr DM, Myers JL, et al. Intraoperative transesophageal echocardiography in a 1.4-kg infant with complex congenital heart disease. Pediatr Cardiol 2003;24:84–5.

[25] Stevenson JG. Incidence of complications in pediatric transesophageal echocardiography: experience in 1650 cases. J Am Soc Echocardiogr 1999;12:527–32.

[26] Kharasch ED, Sivarajan M. Gastroesophageal perforation after intraoperative transesophageal echocardiography. Anesthesiology 2000;85:426–8.

[27] Spahn DR, Schmid S, Carrel T, Pasch T, Schmid ER. Hypopharynx perforation by a transesophageal echocardiography probe. Anesthesiology 1995;82:581–3. Journal of the American Society of Echocardiography Volume 18 Number 1 Ayres et al 97.

[28] Chow MS, Taylor MA, Hanson CWIII. Splenic laceration associated with transesophageal echocardiography. J Cardiol Vasc Anesth 1998;12:314–5.

[29] Savino JS, Hanson CWIII, Bigelow DC, Cheung AT, Weiss SJ. Oropharyngeal injury after transesophageal echocardiography. J Cardiol Vasc Anesth 1994;8:76–8.

[30] Latham P, Hodgins LR. A gastric laceration after transesophageal echocardiography in a patient undergoing aortic valve replacement. Anesth Analg 1995;81:641–2.

[31] Muhiudeen-Russell IA, Miller-Hance WC, Silverman NH. Unrecognized esophageal perforation in a neonate during transesophageal echocardiography. J Am Soc Echocardiogr 2001;14:747–9.

[32] Muhiudeen I, Silverman N. Intraoperative transesophageal echocardiography using high resolution imaging in infants and children with congenital heart disease. Echocardiography 1993;10:599–608.

[33] Stayer SA, Bent ST, Andropoulos DA. Proper probe positioning for infants with compromised ventilation from transesophageal echocardiography. Anesth Analg 2001;92:1073–7.

[34] Madriago EJ, Punn R, Geeter N, Silverman NH. Routine intra-operative trans-oesophageal echocardiography yields better outcomes in surgical repair of CHD. Cardiology Young 2016;26:263–8.

[35] American Society of Anesthesiologists and the Society of Cardiovascular Anesthesiologists Task Force on Transesophageal Echocardiography. Practice guidelines for perioperative transesophageal echocardiography. Anesthesiology 1996;84:986–1006.

[36] Russell IA, Miller-Hance WA, Silverman NH. Intraoperative transesophageal echocardiography for pediatric patients with congenital heart disease. Anesth Analg 1998;87:1058–87.

[37] Bezold LI, Pignatelli R, Altman CA, Feltes TF, Garajski RJ, Vick GW3rd, et al. Intraoperative transesophageal echocardiography in congenital heart surgery. The Texas Children's Hospital experience. Tex Heart Inst J 1996;23:108–15.

[38] Cyran SE, Kimball TR, Meyer RA, Bailey WW, Lowe E, Balisteri WJ, et al. Efficacy of intraoperative transesophageal echocardiography in children with congenital heart disease. Am J Cardiol 1989;63:594–8.

[39] Ungerleider RM. Biplane and multiplane transesophageal echocardiography. Am Heart J 1999;138:612–3.

[40] Stevenson JG. Adherence to physician training guidelines for pediatric transesophageal echocardiography affects the outcome of patients undergoing repair of congenital cardiac defects. J Am Soc Echocardiogr 1999;12:165–72.

[41] Stevenson JG. Performance of intraoperative pediatric transesophageal echocardiography by anesthesiologists and echocardiographers: training and availability are more important than hats. J Am Soc Echocardiogr 1999;12:1013–4.

[42] Meineri M. Transesophageal echocardiography: what the anesthesiologist has to know. Minerva Anestesiologica 2016;82(8):895–907.

[43] Andropoulos DB. Transesophageal echocardiography as a guide to central venous catheter placement in pediatric patients undergoing cardiac surgery. J Cardio Vasc Anesth 1999;13:320–1.

[44] Andropoulos DB, Stayer SA, Bent ST, Campos CJ, Bezold LI, Alvarez M, et al. A controlled study of transesophageal echocardiography to guide central venous catheter placement in congenital heart surgery patients. Anesth Analg 1999;89:65–70.

[45] Fleischer DE, Goldstein SA. Transesophageal echocardiography; what the gastroenterologist thinks the cardiologist should know about endoscopy. J Am Soc Echocardiogr 1990;3:428–34.

[46] Ramamoorthy C, Lynn AM, Stevenson JG. Pro: transesophageal echocardiography should be routinely used during pediatric open cardiac surgery. J Cardiothorac Vasc Anesth 1999;13:629–31.

[47] Reeves ST, Finley AC, Skubas NJ, Swaminathan M, Whitley WS, Glas KE, et al. Basic perioperative transesophageal echocardiography examination: a consensus statement of the American Society of Echocardiography and the Society of Cardiovascular Anesthesiologists. J Am Soc Echocardiogr 2013;26:443–56.

[48] Scohy TV, Ten Cate FJ, Lecomte PV, McGhie J, de Jong PL, Hofland J, et al. Usefulness of intraoperative real-time 3D transesophageal echocardiography in cardiac surgery. J Card Surg 2008;23:784–6.

# FURTHER READING

[49] Matsumoto M, Oka Y, Strom J, Frishman W, Kadish A, Becker RM, et al. Application of transesophageal echocardiography to continuous intraoperative monitoring of left ventricular performance. Am J Cardiol 1980;46:95–105.

[50] Shanewise JS, Cheung AT, Aronson S, Stewart WJ, Weiss, Mark JB, et al. ASE/SCA guidelines for performing a comprehensive intraoperative multiplane transesophageal echocardiography examination: recommendations of the American Society of Echocardiography Council for Intraoperative Echocardiography and the Society of Cardiovascular Anesthesiologists Task Force for Certification in Perioperative Transesophageal Echocardiography. J Am Soc Echocardiogr 1999;12:884–900.

[51] Fanshawe M, Ellis C, Habib S, et al. A retrospective analysis of the costs and benefits related to alterations in cardiac surgery from routine intraoperative transesophageal echocardiography. Anesth Analg 2002;95:824–7.

[52] Singer M. Oesophageal Doppler monitoring: should it be routine for high-risk surgical patients? Curr Opin Anaesthesiol 2011;24:171–6.

[53] Stokes JW. Transoesophageal echocardiography in routine cardiac surgery. Med J Aust 2005;182:93–4.

[54] Scohy TV, Matte G, van Neer PL, et al. A new transesophageal probe for newborns. Ultrasound Med Biol 2009;35:1686–9.

[55] Scohy TV, Gommers D, Schepp MN, et al. Image quality using a micromultiplane transesophageal echocardiographic probe in older children during cardiac surgery. Eur J Anaesthesiol 2009;26:445–7.

[56] Scohy TV, Gommers D, Jan ten Harkel AD, et al. Intraoperative evaluation of micromultiplane transesophageal echocardiographic probe in surgery for congenital heart disease. Eur J Echocardiogr 2007;8:241–6.

[57] Ma XJ, Huang GY, Liang XC, et al. Transoesophageal echocardiography in monitoring, guiding, and evaluating surgical repair of congenital cardiac malformations in children. Cardiol Young 2007;17:301–6.

[58] Balmer C, Barron D, Wright JG, et al. Experience with intraoperative ultrasound in paediatric cardiac surgery. Cardiol Young 2006;16:455–62.

[59] Aronson LA. Transnasal placement of biplane transesophageal echocardiography probe intraoperatively in an adolescent with congenital heart disease. Anesth Analg 2003;97:1617–9.

[60] Rice MJ, McDonald RW, Li X, et al. New technology and methodologies for intraoperative, perioperative, and intraprocedural monitoring of surgical and catheter interventions for congenital heart disease. Echocardiography 2002;19:725–34.

[61] Bruce CJ, O'Leary P, Hagler DJ, et al. Miniaturized transesophageal echocardiography in newborn infants. J Am Soc Echocardiogr 2002;15:791–7.

Chapter 6

# Peripheral Cardiopulmonary Bypass—The Percutaneous Cannulation of the Internal Jugular Vein

Karmi Shafer, Demetrio Pittarello
*University of Padua, Padua, Italy*

## PREOPERATIVE EVALUATION

Examination of all previous surgical and anesthesiological records is necessary. Particular attention should be paid to any difficulties encountered during central venous cannulation. It is important to enquire about possible anatomical anomalies, scars, Port-A-Catheters, and/or pace-maker catheters position that might interfere with the percutaneous cannulation of the internal jugular vein (PCIJV).

The internal jugular vein (IJV) is preferred for venous drainage from the upper side of the body. Cannulation of the right side virtually ensures a central location because the IJV, the superior vena cava and the right atrium are in a straight line. Left-sided cannulation risks injury to the thoracic duct and possible pneumothorax because the apex of the lung is higher on the left. However, when a right IJV access is judged difficult, after more then two failed attempts an alternative access should be considered (left IJV, subclavian vein). A preliminary 2D-echo of the neck is routinely performed to determine vessel's location, size, and its relationship to surrounding structures (Fig. 6.1).

A wrong size jugular venous cannula can cause vena caval obstruction, especially in low-weight children, which may result in cerebral venous congestion and subsequent reduction in cerebral perfusion pressure. In the paediatric population, the size of the venous cannula is crucial and the use of vacuum-assisted venous drainage decreases the diameters of cannulas (usually by 1 to 2 sizes of the estimated size for the patient's body surface area) without compromising the venous return, thus expanding the spectrum of peripheral cannula on to smaller children [1, 2].

## PATIENT PREPARATION

After the induction of general anesthesia, the patient is positioned in slight Trendelenburg position with the head turned 30 degree away from the cannulation side; turning the head too far to the side may result in compression of the vein and moving the vein in closer proximity to the carotid artery (Fig. 6.2).

A rolled towel or sheet is placed under the center of the back to allow slight extension of the head (Fig. 6.3).

Before IJV puncture, a transesophageal (TEE) probe is positioned to assess the final position of the cannula at the superior vena cava-right atrial junction site, once inserted. Usually the best view can be achieved at the mid esophageal position with 90–120 degree visualization angle for the inferior and superior vena cava long axis view also called the bicaval view (Fig. 6.4).

## PCIJV PROCEDURE

The PCIJV procedure is carried-out by using standard sterile eco-guided modified Seldinger technique. Ultrasound-guided venous cannulation has the advantage of leading and facilitating the puncture of the vessel, thereby reducing the risk of complications [3]. The echo-guided puncture of the IJV can be performed by using either the "out-of-plane technique", consisting in the position of the ultrasonic beam just perpendicular to the vessel axis (Fig. 6.5), or the "in-plane technique",

Copyright © 2018 Elsevier Inc. All rights reserved.

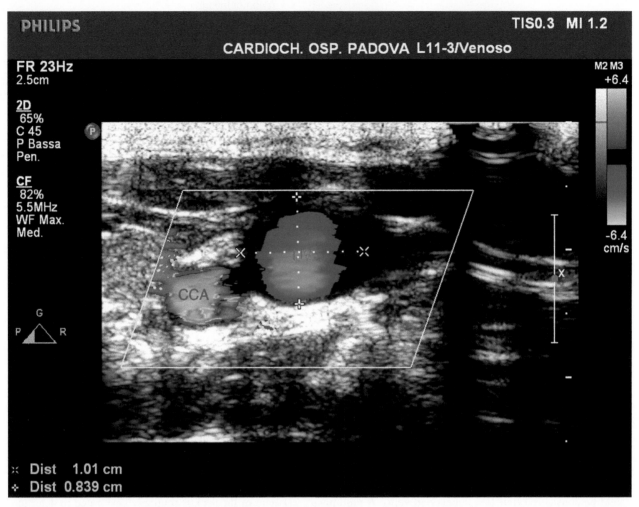

**FIGURE 6.1** **2D-echo scan of the neck made with Philips L11-3 linear ultrasound transducer.** Internal jugular vein (IJV); comune carotide artery (CCA); The image is taken during preliminary 2D-echo scan of the neck, color doppler is added to the evaluation to facilitate recognition and measurement of the vessels. In the lower left corner you can see the sizing of IJV as it was performed by the operator.

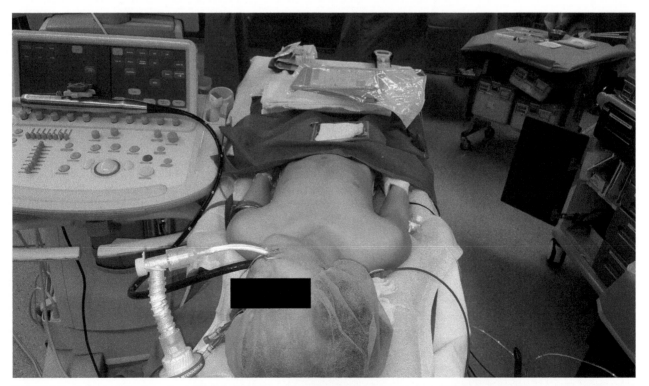

FIGURE 6.2    Patient positioning prior to PCIJV: slight Trendelenburg position, the head is turned 30 degree away from the cannulation side.

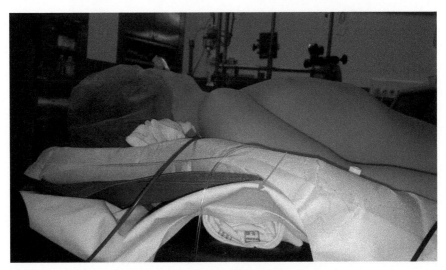

**FIGURE 6.3   Patient positioning, lateral view: Taken during patients preparation to PCIJV.** The *red arrow* indicates the position of rolled sheet placement for slight head extension.

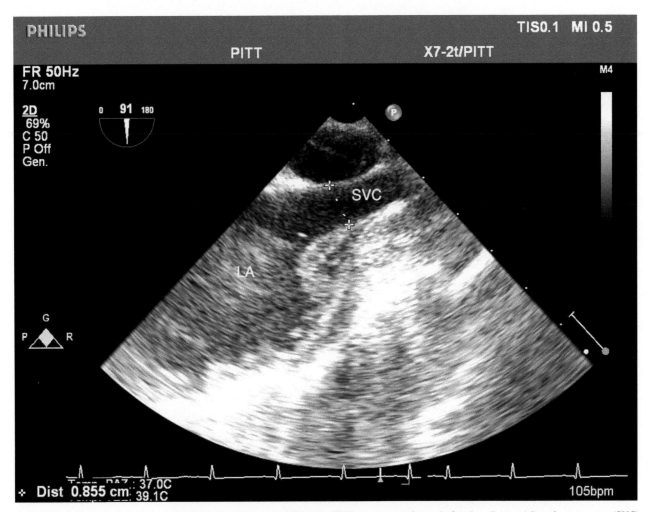

**FIGURE 6.4   Bicaval view made with Philips X7-2t Live 3D TEE xMATRIX array transducer.** Left atrium (LA) and Superior vena cava (SVC) junction in a bicaval view. In the lower left corner you can see SVC final tract diameter as it was measured by the operator. TEE transducer is left in this position for the entire PCIJV procedure for guidance and monitoring.

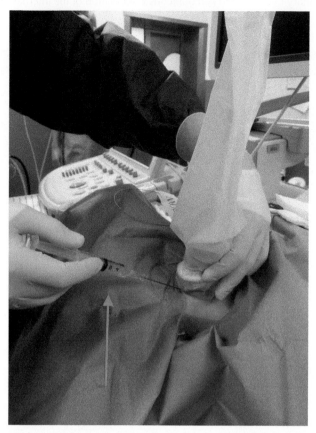

FIGURE 6.5  **In-plane technique visualization using Philips L11-3 linear ultrasound transducer.** Sterile eco-guided modified Seldinger technique, the ultrasonic beam is positioned along the IJV axis. The *red arrow* is indicating patients head position.

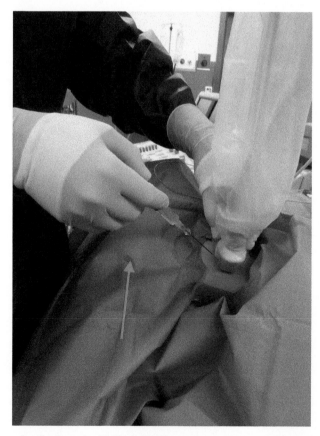

FIGURE 6.6  **Out-of-plane technique visualization using Philips L11-3 linear ultrasound transducer.** Sterile eco-guided modified Seldinger technique, the ultrasonic beam is positioned perpendicular to the IJV axis. The *red arrow* is indicating patients head position.

by positioning the ultrasonic beam along the IJV axis (Fig. 6.6). After IJV puncture, the advancement of the needle along IJV axis is guided by minimal movements ending in the penetration of the anterior wall of the IJV (Fig. 6.7).

Once in the vessel lumen, a guide-wire is carefully advanced under ultrasound guidance (Figs. 6.8 and 6.9).

Subsequently, the progressive dilatation of the IJV percutaneous access is achieved by three dilators, of increasing diameter, to facilitate the introduction of the venous cannula (Figs. 6.10 and 6.11).

After completing the IJV dilatation, the superior venous cannula (Bio-Medicus, Medtronic, United States—which is available in five different sizes, 15 Fr, 17 Fr, 19 Fr, 21 Fr, 23 Fr, and 25 Fr) (Tables 6.1 and 6.2) is inserted and positioned, under TEE-echo guidance, about 1 cm above the junction between superior vena cava and right atrium (Fig. 6.12) and fixed to the skin (Figs. 6.13 and 6.14).

One hundred units per kilogram of sodium heparin are administrated to the patient and the cannula is clamped and flushed with heparinized solution to prevent thrombus formation (Fig. 6.15).

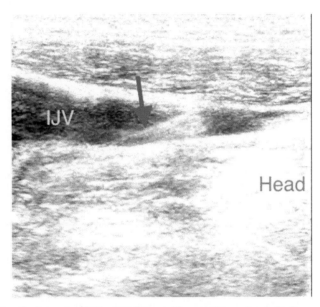

**FIGURE 6.7    In-plane technique visualization using Philips L11-3 linear ultrasound transducer.** Sterile eco-guided modified Seldinger technique, the ultrasonic beam is positioned along the internal jugular vein (IJV) axis. The *red arrow* is indicating the tip of the needle. Patients head is to the right of the image.

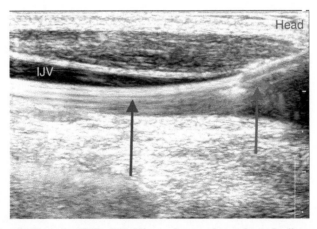

**FIGURE 6.8    In-plane technique visualization using Philips L11-3 linear ultrasound transducer.** Sterile eco-guided modified Seldinger technique, the ultrasonic beam is positioned along the internal jugular vein (IJV) axis. The *red arrow* is indicating the tip of the needle. The *blue arrow* is indicating the guiding wire advancement inside the vessels lumen. Patients head is to the right of the image.

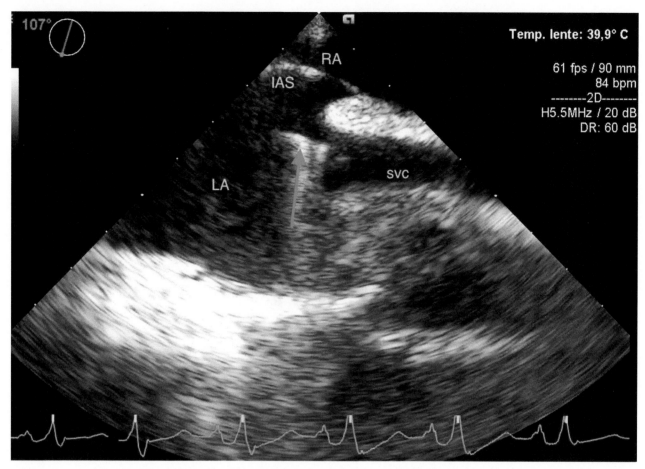

**FIGURE 6.9   Bicaval view made with Philips X7-2t Live 3D TEE xMATRIX array transducer.** Sterile eco-guided modified Seldinger technique. Left atrium (LA) and Superior vena cava (SVC) junction in a bicaval view. In this view you can also see the right atrium (RA) and the interatrial septum (IAS). The *red arrow* is indicating guiding wire advancement into the left atrium.

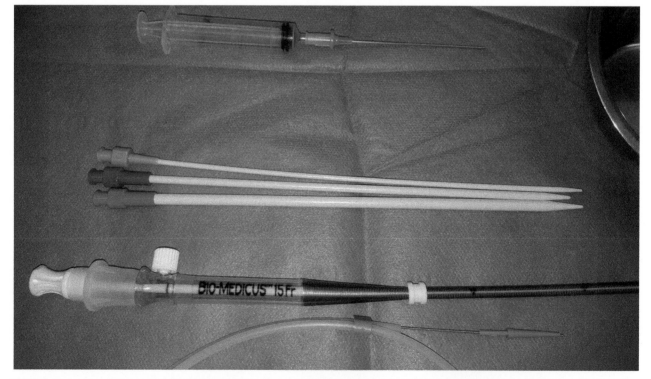

**FIGURE 6.10   Three Dilators and Bio-Medicus Medtronic cannula.** Sterile eco-guided modified Seldinger technique, progressive dilatation of the IJV percutaneous access is achieved by three dilators, of increasing diameter, to facilitate the introduction of the venous cannula.

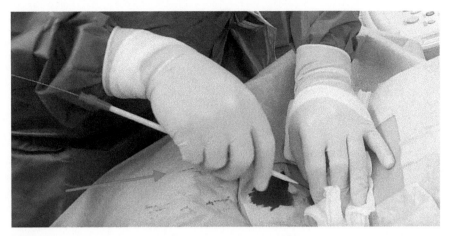

**FIGURE 6.11   Dilatation of the IJV percutaneous access.** Sterile eco-guided modified Seldinger technique, progressive dilatation of the IJV percutaneous access is achieved by three dilators, of increasing diameter, to facilitate the introduction of the venous cannula. The *red arrow* is indicating the head of the patient.

**TABLE 6.1** Author's suggested size of percutaneous IJV cannula for peripheral cardio-pulmonary bypass according to weight group

| IJC cannulas (in French) | Weight group (kg) |
| --- | --- |
| 15 | <35 |
| 17 | 35–65 |
| 21 | >65 |

**TABLE 6.2** Assisted venous drainage: pressure drop versus flow measurement

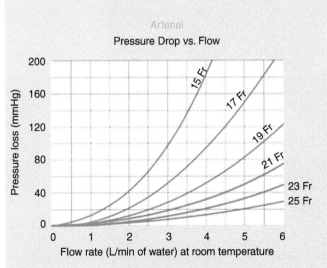

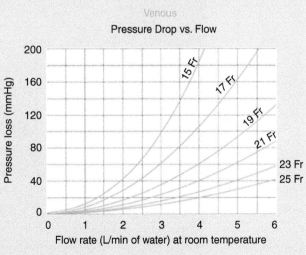

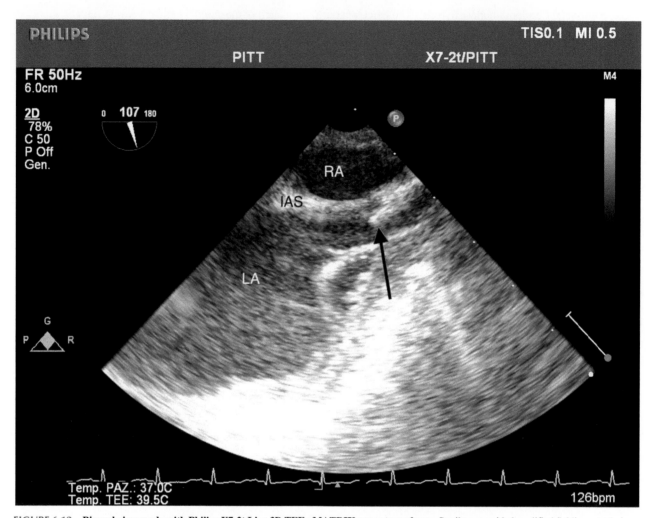

**FIGURE 6.12** **Bicaval view made with Philips X7-2t Live 3D TEE xMATRIX array transducer.** Sterile eco-guided modified Seldinger technique. Left atrium (LA) and Superior vena cava (SVC) junction in a bicaval view. In this view you can also see the right atrium (RA) and the interatrial septum (IAS). The *black arrow* is indicating the cannula advancement toward its final position at the LA-SVC junction.

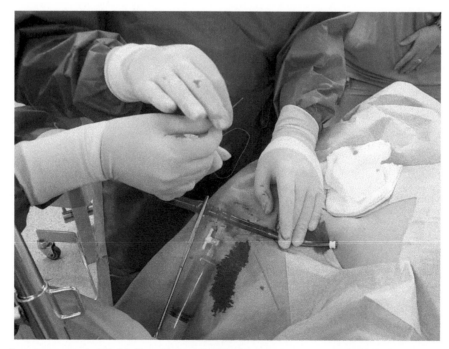

**FIGURE 6.13** **Superior venous cannula (Bio-Medicus, Medtronic, United States).** Superior venous cannula fixated to the skin at insertion point. The *red arrow* is indicating patients head.

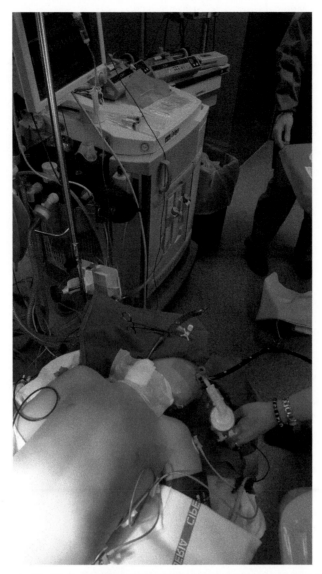

FIGURE 6.14 **Superior venous cannula (Bio-Medicus, Medtronic, United States).** Superior venous cannula fixated to the skin at insertion point.

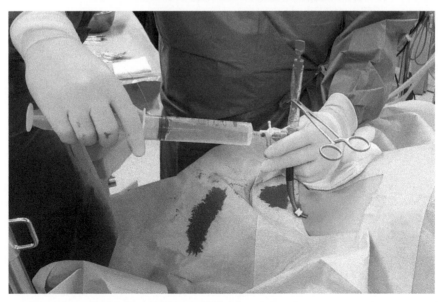

FIGURE 6.15 **Superior venous cannula (Bio-Medicus, Medtronic, United States).** Superior venous cannula is clamped and flushed with heparinized solution to prevent thrombus formation.

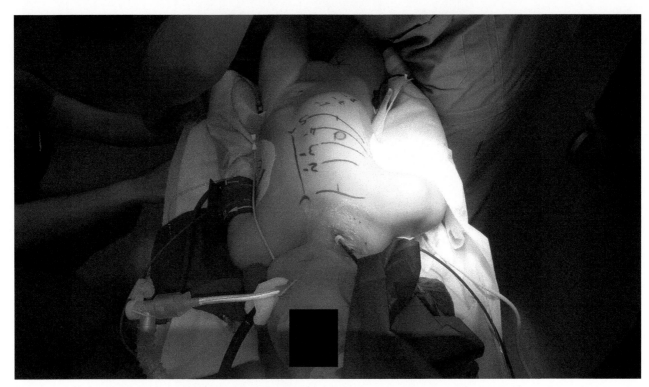

FIGURE 6.16    Patient positioning. 30 degree left lateral decubitus for achieving a right-anterior mini-thoracotomy.

## ON AND OFF PUMP CONDUCT

The patient is then positioned for surgical access (usually a 30 degree left lateral decubitus for achieving a right-anterior mini-thoracotomy or more recently a 0 degree left lateral decubitus for a lateral "axillary" thoracotomy access depending on the congenital heart malformation) (Fig. 6.16). During surgery the cannula needs to remain easily accessible to the anesthesiologist in case it needs to be repositioned under the surgeons request. This is particularly important in patients with right partial anomalous pulmonary venous return at the superior vena cava-right atrial junction, where the tip of the cannula needs to be positioned far away from the anomalous drainage, to allow its surgical correction. Subsequently, additional 200 UI/kg of sodium heparin are given, surgical isolation and cannulation of both femoral artery and vein is performed and the cardiopulmonary bypass (CPB) established.

## IJV DECANNULATION

At the end of the surgical procedure, following complete heparin antagonization with protamine sulfate, the IJV cannula is removed, usually during chest closure. A gentle compression of the IJV cannulation site is maintained for 10–15 min to promote vessel sealing and a medication is eventually used to close the insertion site (Figs. 6.17 and 6.18).

## PATIENT'S SAFETY: NEAR-INFRARED SPECTROSCOPY (NIRS) MONITORING

Regional oxygen saturation measured by means of NIRS is a noninvasive monitoring technique for cerebral and somatic mixed venous oxygen saturation used in clinical practice to detect states of low body perfusion. Similar to pulse-oximetry, NIRS uses the proportionate differential in reflection and absorption of different wavelengths of light (Beer-Lambert Law) to estimate the proportion of hemoglobin saturated with oxygen approximately 2–3 cm below the

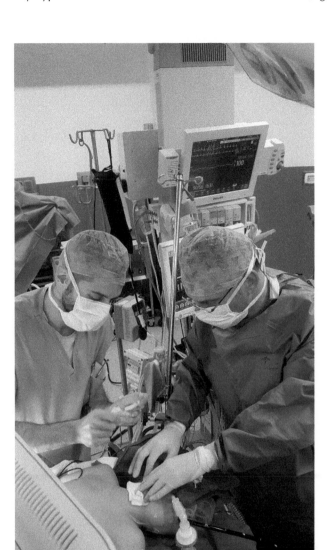

FIGURE 6.17   **Compression of the IJV cannulation site.** A gentle compression of the IJV cannulation site is maintained for 10–15 min to promote vessel sealing.

sensor. The depth of readings is based on the separation between the light source (s) and the receptor (s); the farther the receptor and source are separated, the deeper the arc of penetration. With NIRS, light travels in a banana shaped pathway coming out on the same side of the tissue to sample deep tissue without the need of traveling straight through an extremity. This effect allows NIRS to be used on a broader array of tissues other than just fingers, toes, and ear lobes as in the case of pulse-oximetry. NIRS sampling is predominantly venous (70%–75%) rather than arterial (25%) and does not require pulsatile flow, therefore it is a reliable tissue oxygenation monitoring technique when applied during CPB [4]. Due to baseline levels interpersonal variability, it has been suggested to refer to the NIRS readings variation rather than its absolute value [5, 6]. A 20%–25% drop from baseline level has been associated to postoperative complications and bad outcome [7].

INVOS 5100 cerebral/somatic near infrared spectroscopy oxymeter is used in our center since 2007 to monitor variation of cerebral oxygen saturation during and postsurgery (Fig. 6.19). In patients requiring femoral artery cannulation in addi-

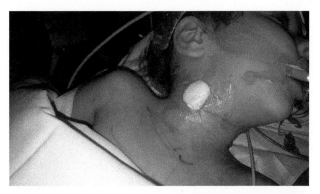

FIGURE 6.18 **Medication of the IJV cannulation site.** A medication is eventually used to close the insertion site.

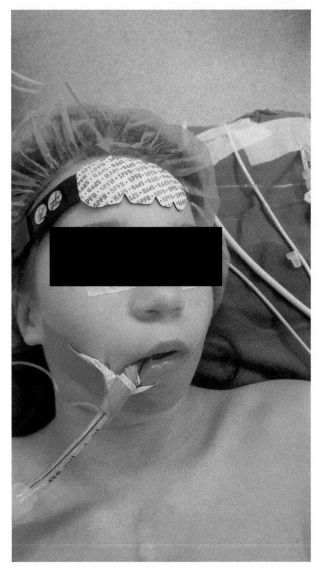

FIGURE 6.19 **Invos Cerebral/Somatic Saturimeter, Somanetics, United States cerebral sensor.** The *red arrow* is indicating the cerebral INVOS sensor position.

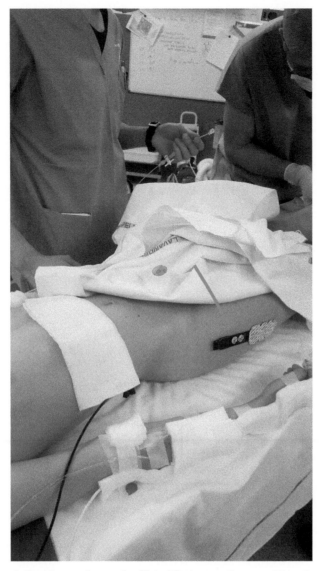

FIGURE 6.20   **Invos Cerebral/Somatic Saturimeter, Somanetics, United States somatic sensor.** The *red arrow* is indicating the somatic INVOS sensor position.

tion to cerebral saturation monitoring, our routine practice is to use NIRS (Invos Cerebral/Somatic Saturimeter, Somanetics, United States) to monitor the cannulated lower extremity during extracorporeal perfusion (Fig. 6.20). The introduction of regional oxygen desaturation score (multiplying the rSO2 <50% of the preoperative baseline value of every single patient, by time in seconds as "rO2 desaturation score = 50% rSo2 − current rSo2 (%) × time in seconds") [10] generated as an area under the curve measurement, which accounts for both depth and duration of desaturation below the 50% saturation threshold [8] and its association with postoperative morbidities [9, 10] has allowed us to identify 50% of basal rSO2 as the lower limit. When somatic NIRS values drop to critical 50% under baseline value during prolonged cardiopulmonary bypass (>80 min) a selective distal perfusion should be considered (Video 6.1). Video 6.1 Peripheral cardio-pulmonary bypass—the percutaneous cannulation of the internal jugular vein.

## REFERENCES

[1] Jegger D, Tevaearai HT, Horisberger J, et al. Augmented venous return for minimally invasive open heart surgery with selective caval cannulation. Eur J Cardiothorac Surg 1999;16:312–6.

[2] Durandy Y. The impact of vacuum-assisted venous drainage and miniaturized bypass circuits on blood transfusion in pediatric cardiac surgery. ASAIO J 2009;55:117–20.

[3] Vida VL, Tiberio I, Gallo M, et al. Percutaneous internal jugular venous cannulation for extracorporeal circulation during minimally invasive technique in children with congenital heart disease: operative technique and results. Minerva Pediatr 2015.

[4] Scheeren TW, Schober P, Schwarte LA. Monitoring tissue oxygenation by near infrared spectroscopy (NIRS): background and current applications. J Clin Monit Comput 2012;26:279–87.

[5] Dullenkopf A, Frey B, Baenziger O, et al. Measurement of cerebral oxygenation state in anaesthetized children using the INVOS 5100 cerebral oximeter. Paediatr Anaesth 2003;13:384–91.

[6] Dullenkopf A, Baulig W, Weiss M, Schmid ER. Cerebral near-infrared spectroscopy in adult patients after cardiac surgery is not useful for monitoring absolute values but may reflect trends in venous oxygenation under clinical conditions. J Cardiothorac Vasc Anesth 2007;21(4):535–9.

[7] Nielsen HB. Systematic review of near-infrared spectroscopy determined cerebral oxygenation during noncardiac surgery. Front Physiol 2014;5:93. Published online 2014 Mar 17.

[8] Slater J, Guarino T, Stack J, et al. Cerebral oxygen desaturation predicts cognitive decline and longer hospital stay after cardiac surgery. Ann Thorac Surg 2009;87:36–45.

[9] Vida VL, Tessari C, Cristante A, et al. The role of regional oxygen saturation using near-infrared spectroscopy and blood lactate levels as early predictors of outcome after pediatric cardiac surgery. Can J Cardiol 2016;32:970–7.

[10] Zulueta JL, Vida VL, Perisinotto E, Pittarello D, Stellin G. Role of intraoperative regional oxygen saturation using near-infrared spectroscopy in the prediction of low output syndrome after pediatric heart surgery. J Card Surg 2013;28:446–52.

## FURTHER READING

[11] Vida VL, Tessari C, Fabozzo A, et al. The evolution of the right anterolateral thoracotomy technique for correction of atrial septal defects: cosmetic and functional results in prepubescent patients. Ann Thorac Surg 2013;95(1):242–7.

[12] Hagl C, Stock U, Haverich A, Steinhoff G. Evaluation of different minimally invasive techniques in pediatric cardiac surgery: is a full sternotomy always a necessity? Chest 2001;119:622–7.

[13] Laussen PC, Bichell DP, McGowan FX, Zurakowski D, DeMasso DR, del Nido PJ. Postoperative recovery in children after minimum versus full-length sternotomy. Ann Thorac Surg 2000;69:591–6.

[14] Nakanishi K, Matsushita S, Kawasaki S, Tambara K, Yamamoto T, Morita T, et al. Morita Tet al Safety advantage of modified minimally invasive cardiac surgery for pediatric patients. Pediatr Cardiol 2013;34:525–9.

[15] Karthekeyan BR, Vakamudi M, Thangavelu P, Sulaiman S, Sundar AS, Kumar SM. Lower ministernotomy and fast tracking for atrial septal defect. Asian Cardiovasc Thorac Ann 2010;18:166–9.

# Chapter 7

# Peripheral Cardiopulmonary Bypass—The Surgical Isolation and Cannulation of Femoral Vessels

Vladimiro L. Vida, Alvise Guariento, Giovanni Stellin
*University of Padua, Padua, Italy*

## INTRODUCTION

Minimally invasive cardiac surgery and closed chest cardiopulmonary bypass techniques are in continuous evolution [1–4]. Saving vital space in less invasive cardiac surgery is of great importance and the introduction of new materials, together with the use of vacuum-assisted venous drainage have allowed a progressively decreasing in size of venous cannula thus making the use of peripheral bypass applicable in patients with progressive lower body weight [5,6].

## CENTRAL VERSUS PERIPHERAL CARDIOPULMONARY BYPASS

According to our minimally invasive surgical protocol, when central cannulation is employed the aortic cannulation is carried on by using a straight aortic cannula (DLP cannulas, Medtronic, United States), and bicaval cannulation can be achieved by using angled venous cannulas (DLP cannulas, Medtronic, United States) or straight armed cannulas (Femflex II, femoral venous cannulation cannula with Duraflo coating, Edward Lifescinces) (Chapter 4).

From June 2008, as an evolution our minimally invasive protocol, we have routinely employed a peripheral cannulation for the cardiopulmonary bypass [7,8] in patients with simple congenital heart defect (CHD) and a body weight superior to 15 kg (this body weight cutoff for peripheral arterial cannulation has been progressively decreased over the last few years, being 35 kg at the beginning of our experience). Remote cardiopulmonary bypass is carried on by percutaneous cannulation of the internal jugular vein (IJV) (described in Chapter 6) followed by surgical isolation and cannulation of the femoral vessels.

## SURGICAL ISOLATION AND CANNULATION OF FEMORAL VESSELS

### Orientation

**H:** head, **F:** foot, **R:** right, **L:** left

### Surgical Preparation and Incision

The patient is prepared by the anesthesiologis in supine position and then turned in left lateral decubitus with the hyperextension of the right thigh on the pelvis (Images 7.1 and 7.2). The groin is marked and the exposure of the femoral artery and vein is obtained through a transversal skin incision (2–3 cms in length) along the groin folder, called "bikini incision" (Images 7.3 and 7.4).

Fundamentals of Congenital Minimally Invasive Cardiac Surgery. http://dx.doi.org/10.1016/B978-0-12-811355-4.00007-1
Copyright © 2018 Elsevier Inc. All rights reserved.

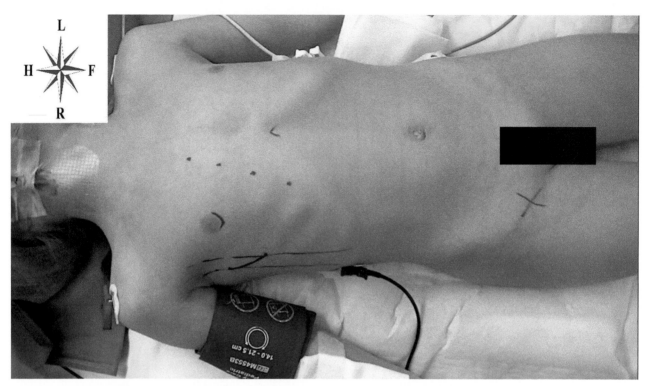

IMAGE 7.1    The patient is prepared by the anesthesiologis in supine position. The incision sites are marked.

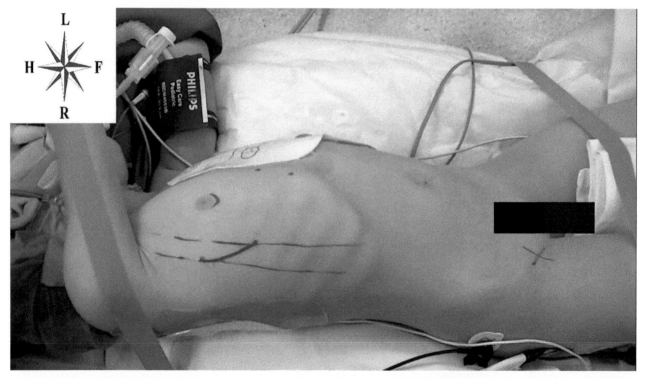

IMAGE 7.2    The patient is then turned in left lateral decubitus with the hyperextension of the right thigh on the pelvis. Note the marked incision sites.

## Femoral Vein and Artery Isolation and Preparation for Cannulation

The femoral incision is speaded with a retractor and the femoral vessels (see Chapter 5) are gently dissected (Images 7.5 and 7.6). Umbilical tapes are the placed around the femoral vein and artery, both proximally and distally to guarantee hemostatic control (Images 7.7–7.10). The femoral arterial and venuous cannulation sites are prepared by placing a "tobacco purse-stitches" (usually a 5.0 polypropylene stitch) in a longitudinal fashion to avoid stenosis at the time of decannulation (Images 7.11–7.13).

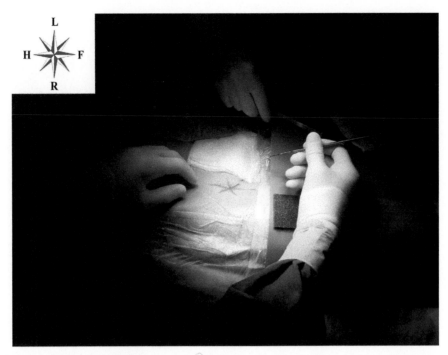

IMAGE 7.3  **The groin is prepared for the incision.**

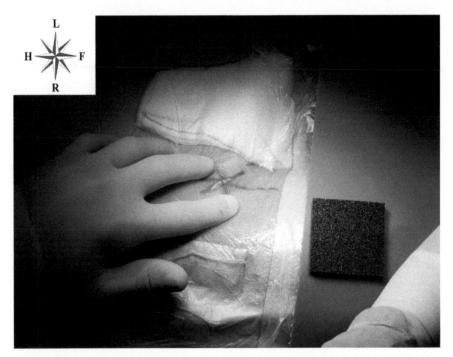

IMAGE 7.4  **Exposure of the femoral artery and vein is obtained through a transversal skin incision (2–3 cm in lenght) along the groin folder, called "bikini incision".**

## Femoral Arterial Cannulation

After complete hemarinization is achieved with a ACT > 400 s, the distal umbilical tape (toward patient's feet) is fixed to the surgical field, while the proximal umbilical tape is controlled by the assistant surgeon (Image 7.14). A small longitudinal incision is made on the anterior surface of the femoral artery along its long axis), within the purse-string by using a blade. The cannulation site is gently enlarged longitudinally with a mosquito clamp (Image 7.15) and the arterial cannula (see Chapter 5) is introduced completely into the lumen of the vessel (till reaching the abdominal aorta-to-ilaic artery junction) (Image 7.16) and it is eventually fixed with a lace (Image 7.17). The cannula is then connected to the cardiopulmonary bypass circuit (with accurate air debubbling) and then further fixed with a stitch to the surgical field to avoid displacement

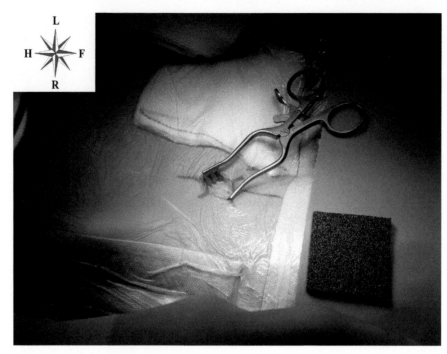

IMAGE 7.5    The femoral incision is speaded with a retractor.

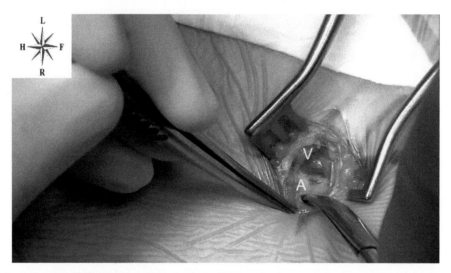

IMAGE 7.6    Femoral vessels are gently dissected. *A*, femoral artery; *V*, femoral vein.

(Image 7.18). In smaller patients (<15 kg of body weight) or in patients with small femoral arteries (smaller in diameter than the arterial cannula to utilize), a central arterial cannulation, into the ascending aorta, is recommended.

## Femoral Vein Cannulation

The same procedure is repeated for the femoral vein. As regards the femoral vein cannula (see Chapter 5), we calculate arbitrarily before inserting the cannula into the femoral vein, the length of the cannula to be inserted (Image 7.19) to reach the inferior vena cava-right atrial junction (which is usually 25 cm in patients with a body weight between 15 and 25 kg, 30 cm in patients with a body weight between 26 and 35 kg and 35–40 cm in bigger patients). A small incision is then made on the surface of the femoral artery, within the purse-string using a blade (Images 7.20 and 7.21). After gentle spreading of the cannulation site (Image 7.22), a Seldinger type introducer (Image 7.23) (see Chapter 5) is introduced in the femoral vein and slided forward under 2D TEE echo guidance until the right atrium is reached (Images 7.24–7.26). The femoral

IMAGE 7.7 Umbilical tapes are the placed around the femoral vein (V) both proximally and distally to guarantee hemostatic control.

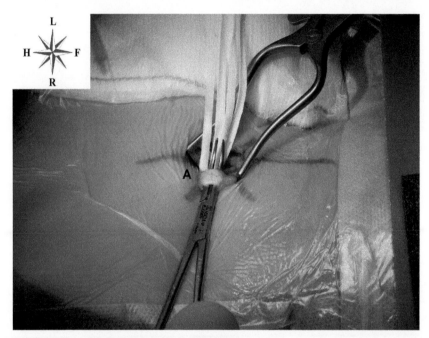

IMAGE 7.8 The femoral artery (A) is gently dissected above the deep femoral artery.

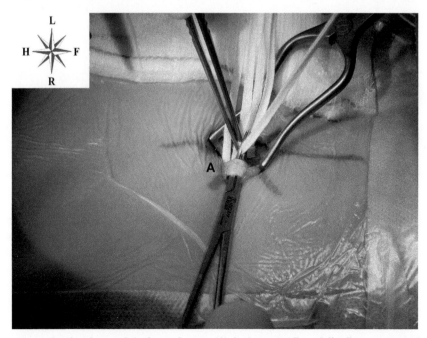

IMAGE 7.9 Umbilical tapes are the placed around the femoral artery (A), both proximally and distally to guarantee hemostatic control.

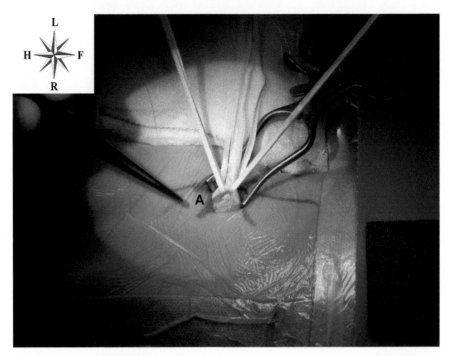

IMAGE 7.10    The femoral artery (A) is gently pulled by using the umbilical tapes to dissect the lower part and to identify collaterals.

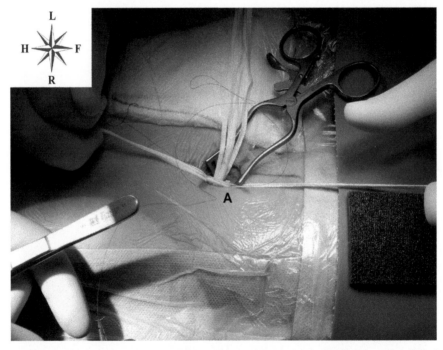

IMAGE 7.11    The femoral arterial cannulation site is prepared by placing a "tobacco purse-stitches" (usually a 5.0 polypropylene stitch) in a longitudinal fashion to avoid stenosis at the time of decannulation. *A*, Femoral artery.

vein cannula is then guided in site by using the Seldinged guide and its final position is decided under 2D echo guidance. Care has to be taken to remove the stilet inside the cannula once it is inserted in the vein to avoid vascular damages. The cannula is eventually fastened with a lace (Image 7.27) and the cardiopulmanary bypass can be initiated (A: femoral artery, V: femorel vein).

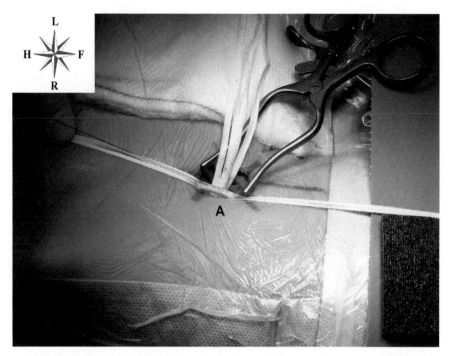

IMAGE 7.12   The femoral arterial "tobacco purse-stitches" usually closed toward the head of the patient. *A*, Femoral artery.

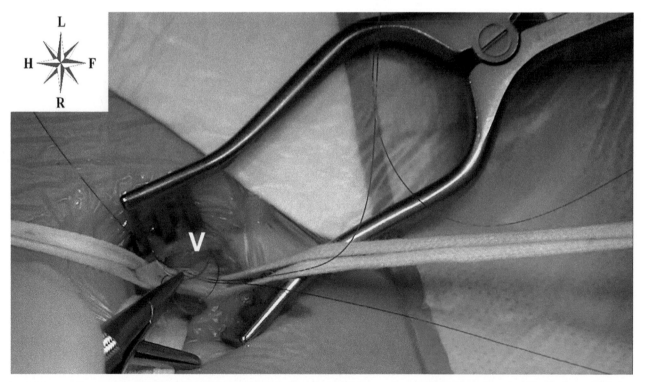

IMAGE 7.13   The femoral venuous cannulation sites isprepared by placing a "tobacco purse-stitches" (usually a 5.0 polypropylene stitch) in a longitudinal fashion to avoid stenosis at the time of decannulation. *V*, Femoral vein

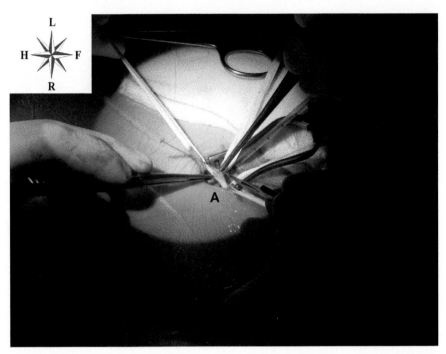

IMAGE 7.14 **Femoral artery cannulation: the distal umbilical tape (toward patient's feet) is fixed to the surgical field, while the proximal umbilical tape is controlled by the assistant surgeon. A small longitudinal incision is made on the anterior surface of the femoral artery along its long axis), within the purse-string by using a blade. The cannulation site is gently enlarged longitudinally with a mosquito clamp. *A,* Femoral artery.**

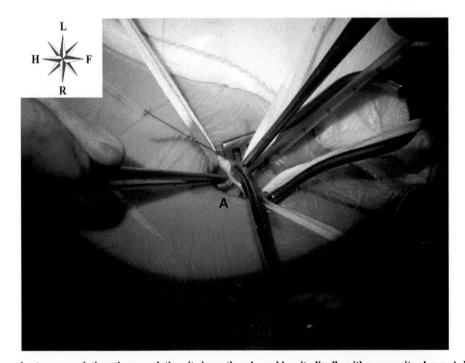

IMAGE 7.15 **Femoral artery cannulation: the cannulation site is gently enlarged longitudinally with a mosquito clamp. *A,* Femoral artery.**

## Femoral Vein and Artery Decannulation

At the end of the cardiopulmonary bypass, after protamin sulfate is administered to neutralize systemic heparinization, decannulation is performed. Femoral decannulation takes place in two phases: the first for the vein followed by the one for the artery. For the femoral vein, umbilical tapes are stretched to reduce venous return (Image 7.28), the venous cannula is

IMAGE 7.16   **Femoral artery cannulation: the arterial cannula is introduced completely into the lumen of the vessel (till reaching the abdominal aorta-to-ilaic artery junction). *A*, Femoral artery.**

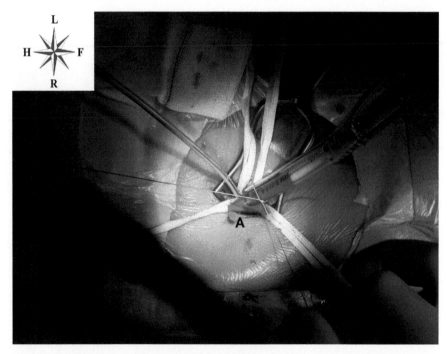

IMAGE 7.17   **Femoral artery cannulation: the arterial cannula (is eventually fixed with a lace. *A*, Femoral artery.**

then retracted (Image 7.29) and the tobacco purse stitch is tied (Image 7.30). For the femoral artery, two vascular clamps are utilized, one distally (Image 7.31) and one proximally (Image 7.32) to the cannulation site; the last one is positioned once the arterial cannula is removed (Image 7.33). The purse string is eventually tied under reduced vascular tension (Image 7.34). In case of residual bleeding through the cannulation site, additional stitches (usually 6.0 polypropylene sutures)

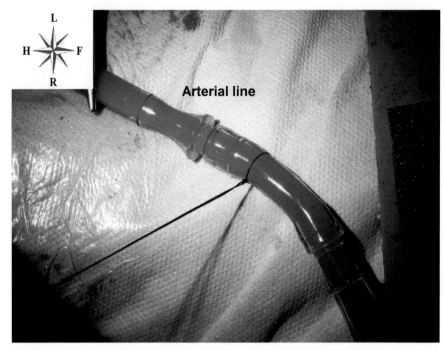

IMAGE 7.18 Femoral artery cannulation: the arterial cannula is connected to the cardiopulmonary bypass circuit (with accurate air debubbling) and then further fixed with a stitch to the surgical field to avoid displacement. *A*, Femoral artery.

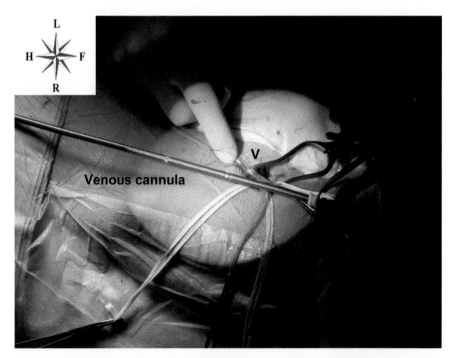

IMAGE 7.19 Femoral vein cannulation: as regards the femoral vein cannula, we calculate arbitrarily before inserting the cannula into the femoral vein, the length of the cannula to be inserted to reach the inferior vena cava-right atrial junction. *V*, femoral vein.

IMAGE 7.20    Femoral vein cannulation: the distal umbilical tape (toward patient's feet) is fixed to the surgical field, while the proximal umbilical tape is controlled by the assistant surgeon. A small longitudinal incision is made on the anterior surface of the femoral vein along its long axis), within the purse-string by using a blade. *V*, Femoral vein.

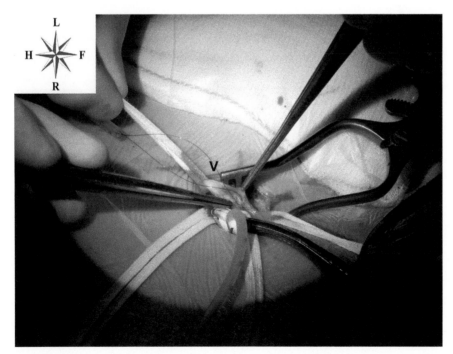

IMAGE 7.21    Femoral vein cannulation: the femoral vein incision is gently opened. *V*, Femoral vein.

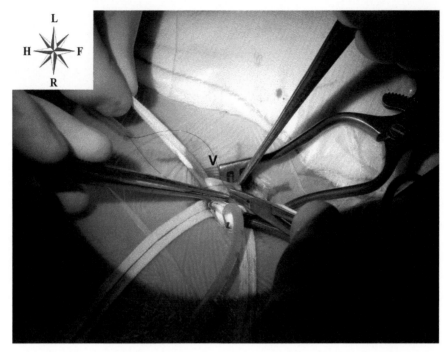

IMAGE 7.22    Femoral vein cannulation: The femoral vein incision is gently spreaded by using a mosquito clamp. *V*, Femoral vein.

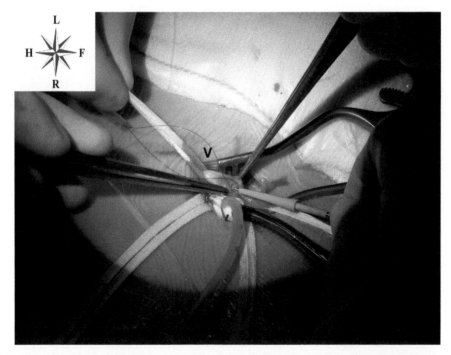

IMAGE 7.23    Femoral vein cannulation: a Seldinger type introducer is introduced in the femoral vein. *V*, femoral vein.

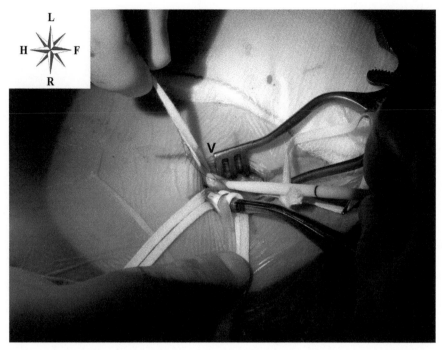

IMAGE 7.24   Femoral vein cannulation: the femoral vein cannula is inserted into the femoral vein. *V*, Femoral vein.

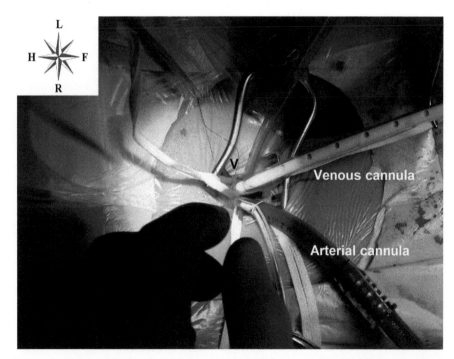

IMAGE 7.25   Femoral vein cannulation: the femoral vein cannula is then slided forward under 2D TEE echo guidance until the right atrium is reached. *V*, Femoral vein.

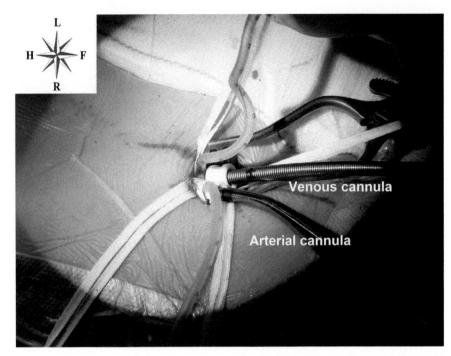

IMAGE 7.26    Femoral vein cannulation: care has to be taken to remove the stilet inside the cannula once it is inserted in the vein to avoid vascular damages.

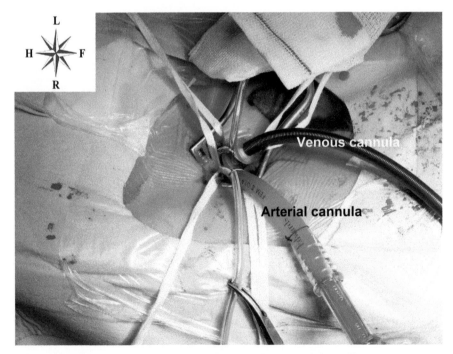

IMAGE 7.27    Femoral vein cannulation: the cannula is eventually fastened with a lace and the cardiopulmanary bypass can be initiated.

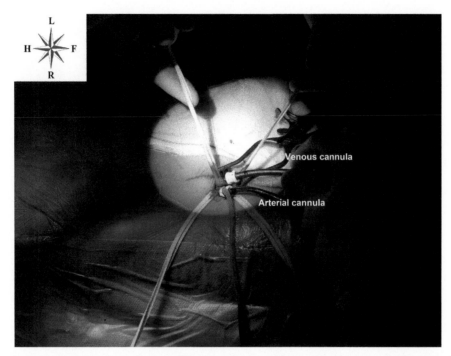

IMAGE 7.28   Femoral vein decannulation: femoral vein, umbilical tapes are stretched to reduce venous return. The femoral artery, two vascular clamps are utilized, one distally (Image 7.31) and one proximally (Image 7.32) to the cannulation site; the last one is positioned once the arterial cannula is removed (Image 7.33). The purse string is eventually tied under reduced vascular tension (Image 7.34)

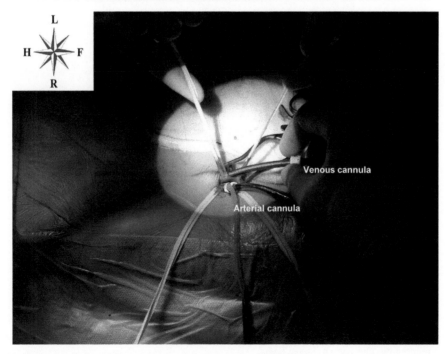

IMAGE 7.29   Femoral vein decannulation: the venous cannula is retracted and pulled out.

IMAGE 7.30   Femoral vein decannulation: the tobacco purse stitch on the femoral vein is tied.

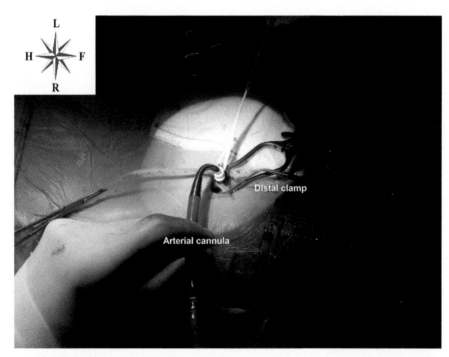

IMAGE 7.31   Femoral artery decannulation: a vascular clamps is placed on the femoral artery, distally to the cannulation site. The purse string is eventually tied under reduced vascular tension (Image 7.34).

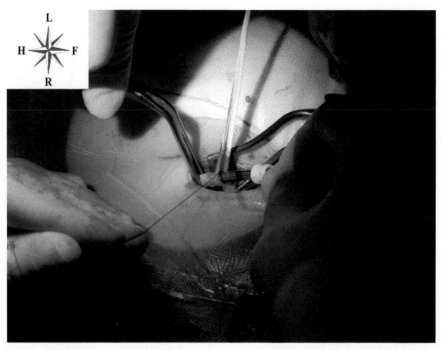

**IMAGE 7.32   Femoral artery decannulation: a vascular clamps is placed on the femoral artery, proximally to the cannulation site once the arterial cannula is removed**

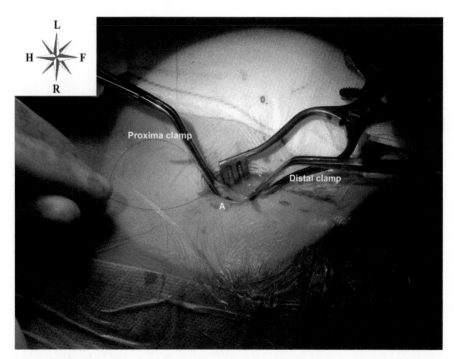

**IMAGE 7.33   Femoral artery decannulation: the femoral arterial cannula heve been pulled out. The two vascular clamps are in site to prevent blood loss.**

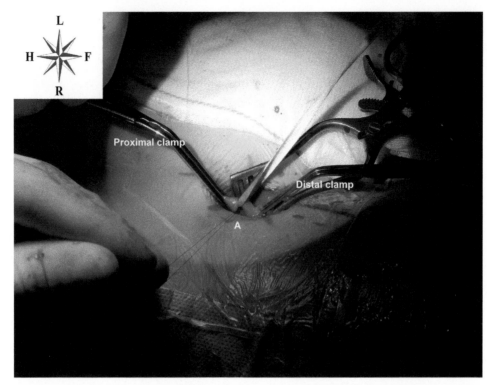

IMAGE 7.34    Femoral artery decannulation: the purse string on the femoral artery is tied under reduced vascular tension. *A*, Femoral artery.

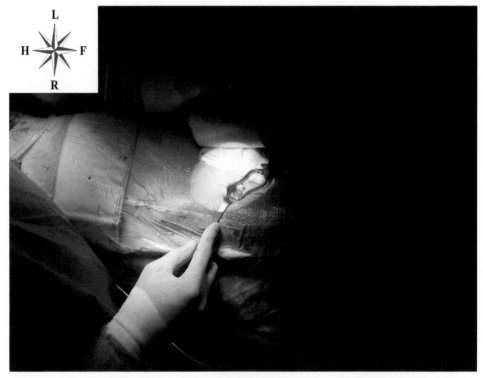

IMAGE 7.35    Femoral decannulation: at the end of the procedure, careful hemostasis is performed.

are positioned to achieve a complete hemostasis. In patients with smaller femoral arteries or in case of suspected arterial stenosis following decannulation, an autologous pericarial patch augmentation of the cannulation site is recommended. At the end of the procedure, careful hemostasis is performed (Images 7.35 and 7.36), and distal pulsation of the femoral artery is monitored directly and by means of NIRS (Image 7.37; Video 7.1 Surgical isolation and cannulation of femoral vessels.

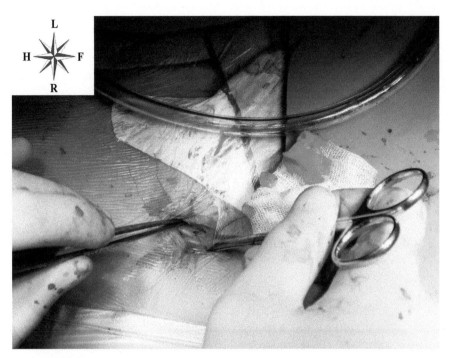

IMAGE 7.36   Femoral decannulation: distal pulsation of the femoral artery is monitored directly.

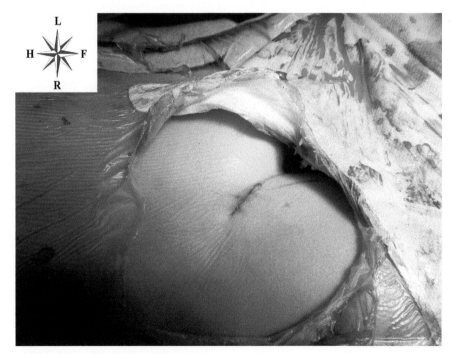

IMAGE 7.37   Femoral incision at the end of the procedure.

## REFERENCES

[1] Cremer JT, Böning A, Anssar MB, et al. Different approaches for minimally invasive closure of atrial septal defects. Ann Thorac Surg 1999;67(6):1648–52.

[2] Vida VL, Padalino MA, Motta R, Stellin G. Minimally invasive surgical options in pediatric heart surgery. Expert Rev Cardiovasc Ther 2011;9(6):763–9.

[3] Vida VL, Padalino MA, Bhattarai A, Stellin G. Right posterior-lateral minithoracotomy access for treating congenital heart disease. Ann Thorac Surg 2011;92(6):2278–80.

[4] Vida VL, Tessari C, Fabozzo A, Padalino MA, Barzon E, Zucchetta F, et al. The evolution of right anterior-lateral thoracotomy technique for correction of atrial septal defects: cosmetic and functional results in pre-pubescent patients. Ann Thorac Surg. 2013;95(1):242–7.

[5] Muhs BE, Galloway AC, Lombino M, et al. Arterial injuries from femoral artery cannulation with port access cardiac surgery. Vasc Endovascular Surg 2005;39(2):153–8.

[6] Vida VL, Tessari C, Putzu A, Tiberio I, Guariento A, Gallo M, et al. The peripheral cannulation technique in minimally invasive congenital cardiac surgery. Int J Artif Organs 2016;19;39(6):300–3.

[7] Vida VL, Padalino MA, Boccuzzo G, Stellin G. Near-infrared spectroscopy for monitoring leg perfusion during minimally invasive surgery for patients with congenital heart defects. J Thorac Cardiovasc Surg 2012;143(3):756–7. Mar.

[8] Vida VL, Tessari C, Putzu A, Tiberio I, Guariento A, Gallo M, et al. The peripheral cannulation technique in minimally invasive congenital cardiac surgery. Int J Artif Organs 2016;19;39(6):300–3. Aug.

# Chapter 8

# Midline Lower Mini-Sternotomy (MS)

Vladimiro L. Vida, Alvise Guariento, Giovanni Stellin
*University of Padua, Padua, Italy*

The midline lower mini-sternotomy approach has been traditionally employed to correct simple CHD lesions as: atrial septal defects, partial atrio-ventricular septal defects and ventricular septal defects [1–4]. However the application of this technique has also been recently extended to correct selected patients with other more complex CHD as complete AV canal defects, tetralogy of Fallot, subaortic stenosis, and so on.

At the beginning of our experience a 6–7 cm skin incision in the midline of the chest had been adopted, with divisions being the apophisis xifoides and the body of the sternum. Later on (since 2007), due to the optimization of the retraction system (Bookwalter retractor, Codman, Germany) and the use of vacuum-assisted venous drainage (with consequent reduction of the caliber of the venous cannula, a 3–4 cm skin incision in the midline of the chest is usually employed, starting about 2 cm below the nipple level. The sternum is longitudinally divided only in its lower third (only the apophisis xifoides or 1–2 cm of the body) and retracted and lifted to expose the aorta.

Due to the small surgical field, to optimize surgical maneuvers, the inferior vena cava cannula is usually inserted in the chest through a separate 5 mm skin incision below the main midline incision (site which is later utilized for inserting a chest drainage at the end of the operation).

The right pleura is routinely opened and the pericardium is incised laterally down to 1 cm from the right phrenic nerve for avoiding possible cardiac tamponade due to postoperative pericardial effusion (now draining into the pleural space); this has found to be related also to the decrease incidence of postcardiotomy syndrome. Sternal closure is usually achieved with interrupted resorbable sutures (usually 2.0 or 3.0) in patients with body weight less than 10 kg and with metal sutures in bigger patients.

In older patients (with a body weight above 20 kg) where the sternum is cut longitudinally in its lower third a transversal T-shape bode incision is performed. This is because the sternum in these patients is completely ossified, and cannot be retracted appropriately. Following the T-shape sternal incision the sternus is retracted and lifted with the Bookwalter retractor. Sternal closure is achieved by using metal stitches.

## SURGICAL TECHNIQUE

### Orientation

**H:** head, **F:** foot, **R:** right, **L:** left.

### Patient Preparation and Midline Mini-Sternotomy

The patient is placed in a supine position. The sternum's structures (manubrium, body, and xiphoid process) are identified, as well as the first six intercostal spaces. Particular attention is taken in positioning the sternum horizontally with the aid of a roll-shaped support placed under the shoulders, at the scapular level. In this way the most cranial structures stand up once the mediastinum is opened, thus facilitating the access to the great vessels (Image 8.1).

The skin incision is limited to as short as 3–4 cm, starting 1–2 cm below the transverse mammillary line (Images 8.2–8.7). It is obviously very important that the incision results precisely in the midline. The pleura are taken off from their sternal

Copyright © 2018 Elsevier Inc. All rights reserved.

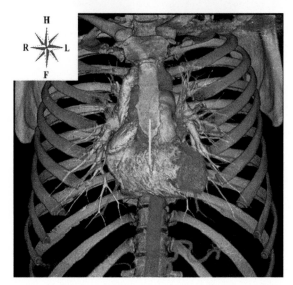

IMAGE 8.1 Angio-CT scan 3D reconstruction in a control patient showing the relationship between the sternum and the anterior mediastinal structures. *Yellow* line showing the position of the midline skin incision. A: aorta, P: main pulmonary artery trunk, RA. Right atrium, RV: right ventricle.

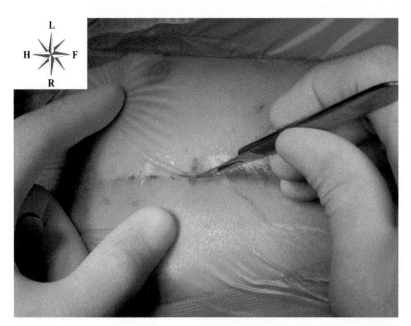

IMAGE 8.2 The patient is prepared in supine position and the skin is marked.

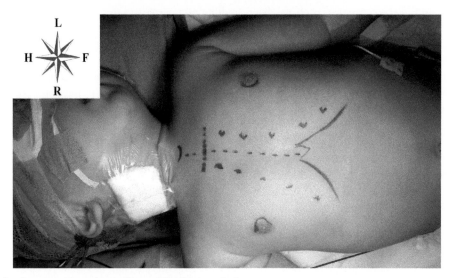

IMAGE 8.3 The skin incision is limited to as short as 3–4 cm, starting 1–2 cm below the transverse mammillary line. It is obviously very important that the incision results precisely in the midline.

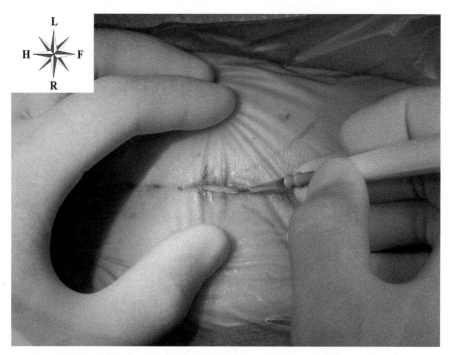

IMAGE 8.4   Electrocautery is used for dissection.

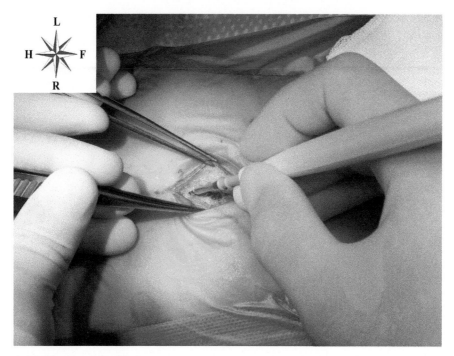

IMAGE 8.5   The sternal xiphoid process is identified.

reflections by using a straight scissor (Images 8.8 and 8.9). The sternal incision (usually made with a "Schumacher sternum shear" (see Chapter 4) (Images 8.10 and 8.11) in patients with a body weight less than 15 kg) is limited to the lower third of the sternum. Electrocautery is used for dissection of the sternal incision (Image 8.12).

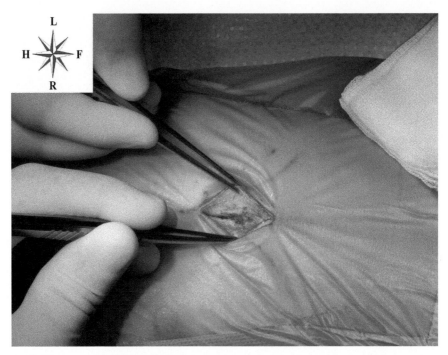

IMAGE 8.6   The sternal xiphoid process is incided with electrocautery.

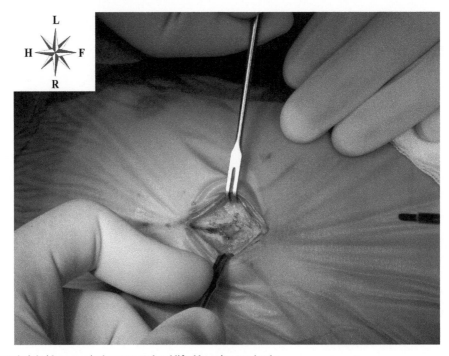

IMAGE 8.7   The sternal xiphoid process is than retracted and lifted by using two hooks.

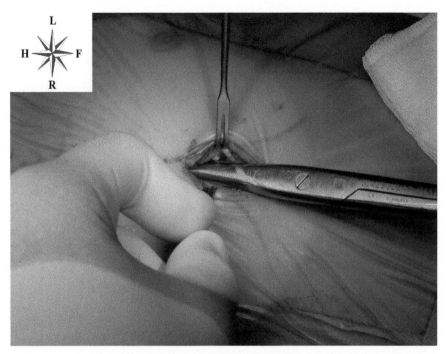

IMAGE 8.8   The pleura are taken off from their sternal reflections by using a straight scissor.

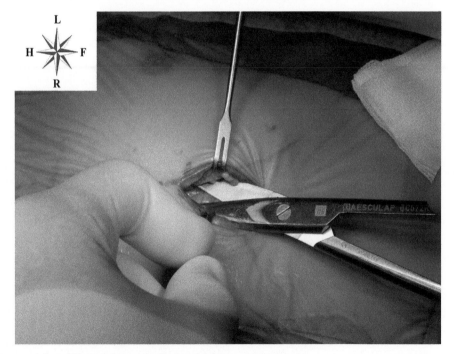

IMAGE 8.9   The pleura are taken off from their sternal reflections by using a straight scissor.

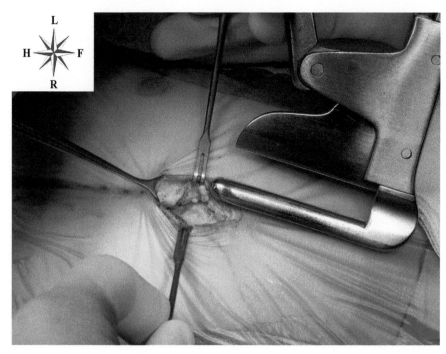

IMAGE 8.10  The sternal incision is usually made with a "Schumacher sternum shear" (see Chapter 4) in patients with a body weight less than 15 kg).

IMAGE 8.11  The sternal incision is then limited to the lower third of the sternum.

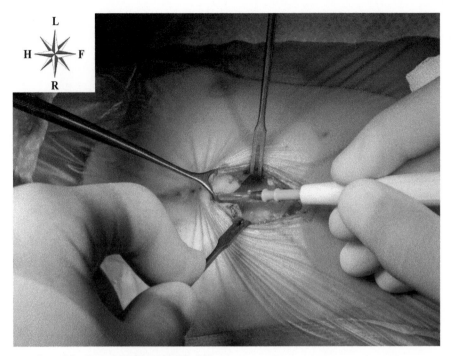

IMAGE 8.12    Electrocautery is used for dissection of the sternal incision.

IMAGE 8.13    A proper exposure is achieved by traction carried out with a sternal retractor (Finochietto retractor).

## Sternal Retraction

A proper exposure is achieved by traction carried out with a sternal retractor (Finochietto retractor) (Image 8.13) (see Chapter 4) and a "Farabeuf" double-ended retractor attached to the Bookwalter Retractor System (Symmetry Surgical Inc., Antioch, TN, USA), which is used to lift and raise the sternum (Images 8.14 and 8.15).

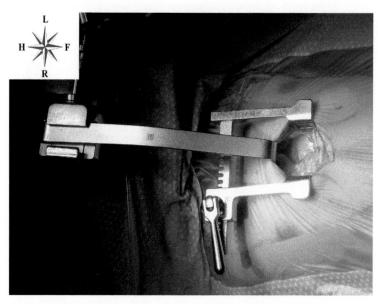

IMAGE 8.14 A "Farabeuf" double-ended retractor is attached to the Bookwalter Retractor System (see Chapter 4), which is used to lift and raise the sternum.

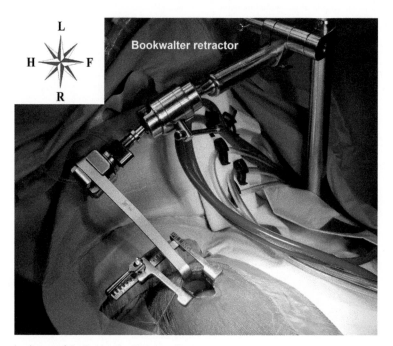

IMAGE 8.15 Intraoperative image of the Bookwalter Retractor System.

## Thymectomy

Partial or subtotal thymectomy is routinely performed to facilitate the visualization of the anterior mediastinal structures (Images 8.16–8.18).

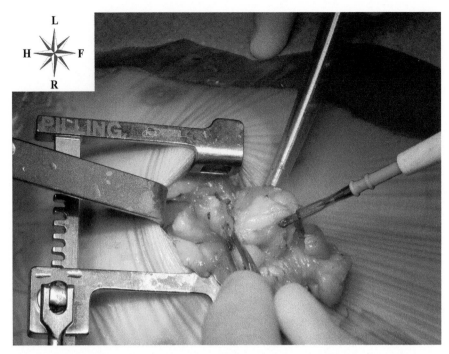

IMAGE 8.16   Partial or subtotal thymectomy is routinely performed to facilitate the visualization of the anterior mediastinal structures.

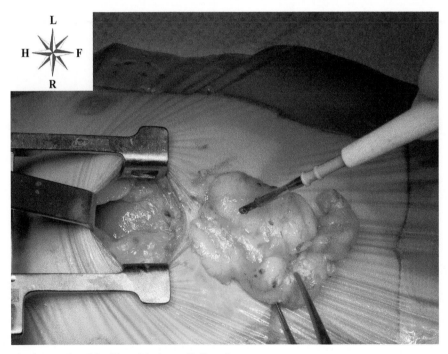

IMAGE 8.17   The proximal extremity of the thimus lobe is usually ligated.

IMAGE 8.18    The visualization of the anterior mediastinal structures is enhanced after thimus removal.

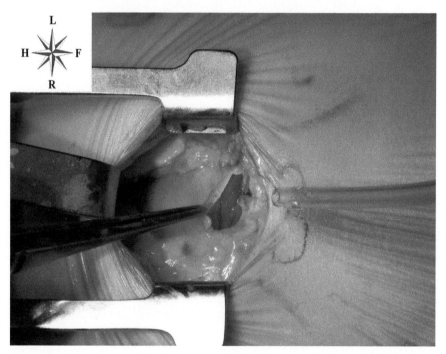

IMAGE 8.19    An adequate portion of the pericardium is then harvested and subsequently fixed in glutaradehide 6% for 10 min.

## Pericardial Patch Harvesting

An adequate portion of the pericardium is then harvested (Images 8.19–8.25) and subsequently fixed in glutaradehide 6% for 10 min. The remaining part of the pericardium is then suspended outside the field to expand the operative view (Images 8.26–8.28).

## Purse String and Cannulation

A straight vascular clamp is placed on the base of the right atrial appendage and a purse string (usually 5.0 polypropylene suture) is placed to facilitate venous cannulation (Images 8.29–8.30). A longitudinal diamond-shaped tobacco purse-string is done in the ascending aorta close to the cephalo-brachial vessels. After full heparinization aortic cannulation is achieved

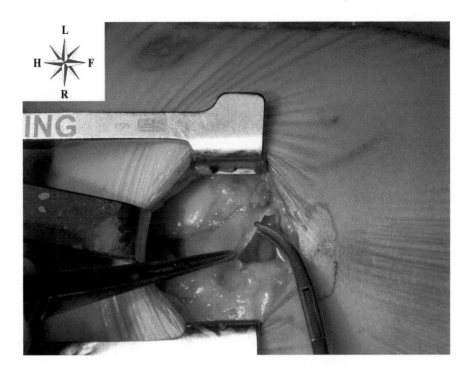

IMAGE 8.20

**IMAGES 8.20 and 8.21**   An adequate portion of the pericardium is then harvested and subsequently fixed in glutaradehide 6% for 10 min.

(Images 8.31 and 8.32). The cannula is then adjusted at an appropriate depth (usually 1 cm) and then fixed with a lace (Image 8.33). A single angled (metal tip) cannula is inserted into the right atrial appendage and the CPB is started (Image 8.34).

A longitudinal tobacco purse-stringing is done cranial to the superior vena cava-to-right atrial junction and the superior vena cava (SVC) is then cannulate (Images 8.35–8.38).

A tobacco purse-string is placed at the level of the interatrial groove to facilitate the insertion of a left atrial suction cannulate (LA (left atrium) vent) (Images 8.39–8.42).

IMAGE 8.22

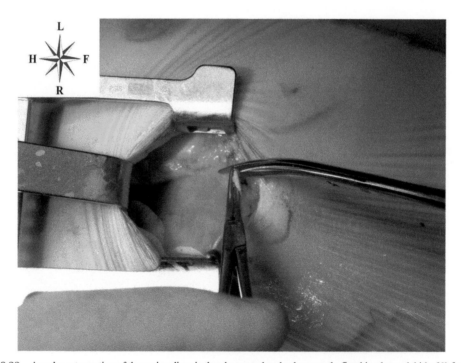

IMAGES 8.22 and 8.23    An adequate portion of the pericardium is then harvested and subsequently fixed in glutaradehide 6% for 10 min.

The Bookwalter retractor is then released to have a better visualization on the inferior vena cava (IVC) zone. A diamond shape tobacco purse-string is done and the IVC is cannulated. The IVC cannula is usually passed through a separate small chest incision (0.5 cm, caudally to the main chest incision), where the thoracic drainage tubes will be positioned at the end of the operation (Images 8.43–8.48).

Vessel loops are placed around SVC and IVC. A Satinsly clamp (see Chapter 4) is used for encircling the IVC (Image 8.49) and a right angle clamp for the SVC (Image 8.50).

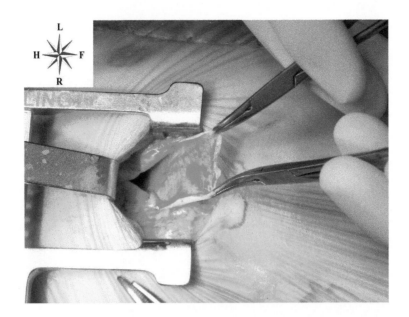

IMAGE 8.24

IMAGES 8.24 and 8.25    An adequate portion of the pericardium is then harvested and subsequently fixed in glutaradehide 6% for 10 min.

IMAGE 8.26    The remaining part of the pericardium is then suspended outside the field to expand the operative view (diaphragmatic pericardium).

IMAGE 8.27   The remaining part of the pericardium is then suspended outside the field to expand the operative view (left pericardium).

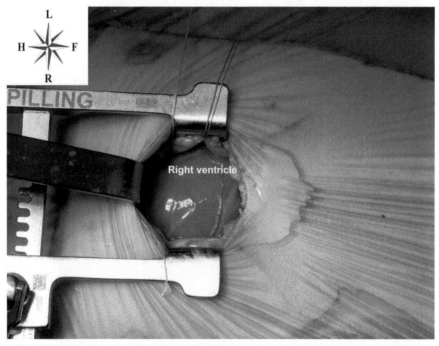

IMAGE 8.28   Operative vision of the heart after suspection of the pericardium.

The Bookwalter retractor is then repositioned and a cardioplegia needle is placed in the ascending aorta (Image 8.51) following by the positioning of the aortic cross clamp and the induction of a cardioplegic arrest (Images 8.52 and 8.53).

The Bookwalter retractor is then again released to facilitate the vision of the right atrium and surgical correction is routinely performed (A: aorta).

Right atriotomy is performed with visualization of the right atrial structure (a pump sucker is visible within the atrial septal communication) (Image 8.54). The ventricular septal defect is eventually closed by using the autologous pericardial patch with a tunning 6.0 polypropylene suture (Image 8.55).

The right atrial incision in then sutured (Image 8.56).

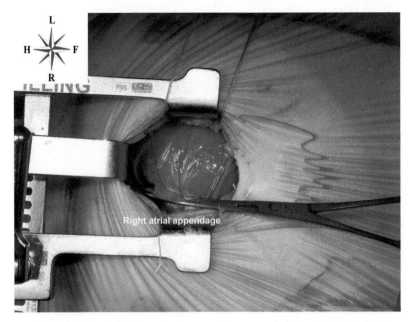

IMAGE 8.29   A straight vascular clamp is placed on the base of the right atrial appendage and the top part is excised.

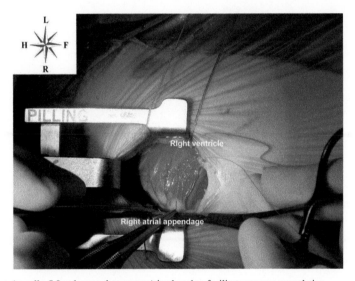

IMAGE 8.30   A purse string (usually 5.0 polypropylene suture) is placed to facilitate venous cannulation.

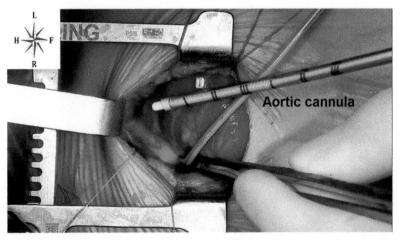

IMAGE 8.31   A longitudinal diamond-shaped tobacco purse-string is done in the ascending aorta close to the cephalo-brachial vessels.

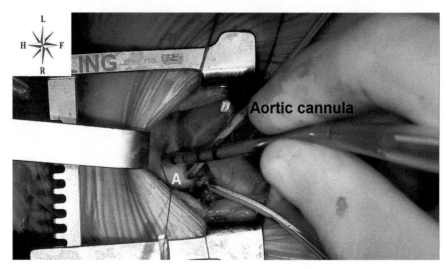

IMAGE 8.32  After full heparinization aortic cannulation is achieved.

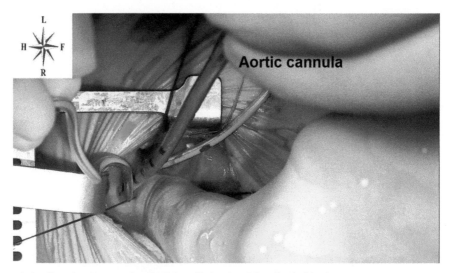

IMAGE 8.33  The cannula is adjusted at an appropriate depth (usually 1 cm) and then fixed with a lace.

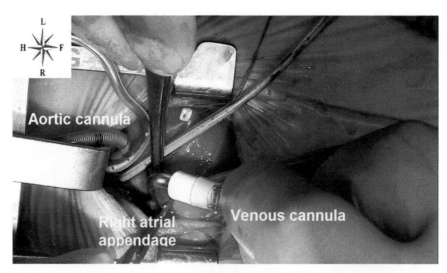

IMAGE 8.34  A single angled (metal tip) cannula is inserted into the right atrial appendage and the CPB is started.

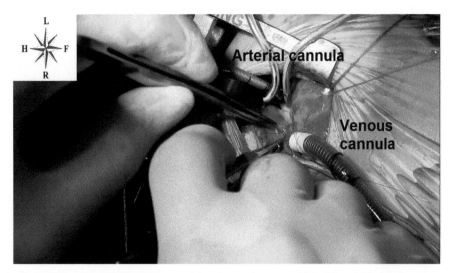

IMAGE 8.35   A longitudinal tobacco purse-stringing is done cranial to the superior vena cava-to-right atrial junction.

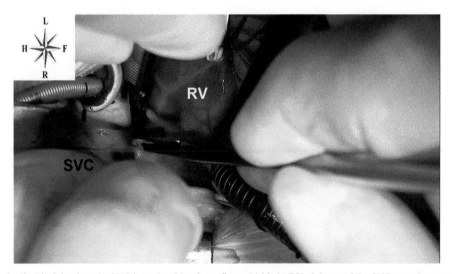

IMAGE 8.36   A longitudinal insizion into the SVC is made with using a figure-11 blade. RV, right ventricle; SVC, superior vena cava.

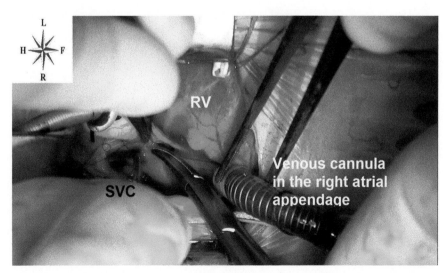

IMAGE 8.37   The incision on the SVC is gently spreaded with a mosquito clamp. RV, right ventricle; SVC, superior vena cava.

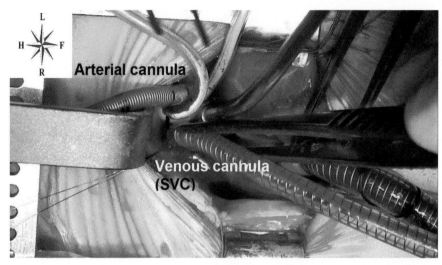

IMAGE 8.38   SVC cannulation is achieved (usually by using a forcep or a mosquito clamp for the limited space). SVC, superior vena cava.

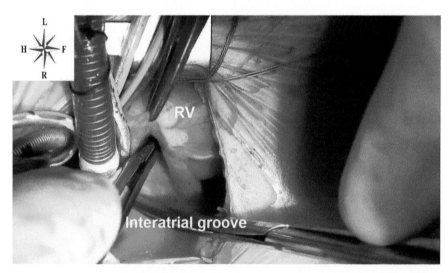

IMAGE 8.39   The interatrial groove is identified and disscted. RV, right ventricle.

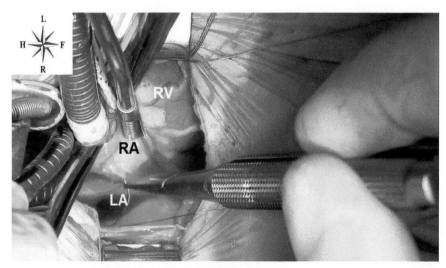

IMAGE 8.40   A tobacco purse-string is placed at the level of the interatrial groove to facilitate the insertion of a left atrial suction cannulate. LA, left atrium; RA, right atrium; RV,  right ventricle.

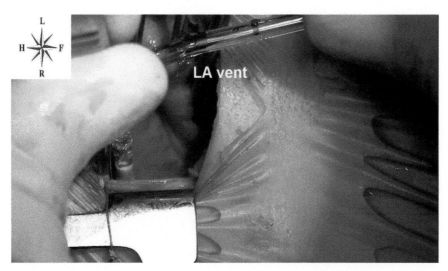

IMAGE 8.41    The LA vent is inserted. LA, left atrium

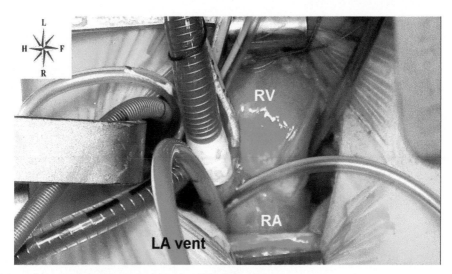

IMAGE 8.42    The LA vent is on. LA, left atrium; RA, right atrium; RV, right ventricle.

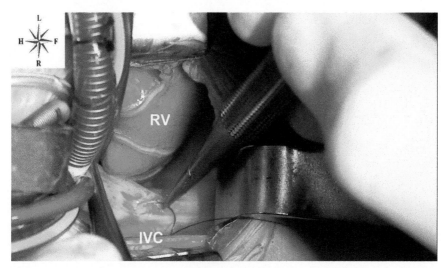

IMAGE 8.43    The Bookwalter retractor is then released to have a better visualization on the inferior vena cava (IVC) zone. A diamond shape tobacco purse-string is done and the IVC is cannulated. IVC, inferior vena cava; RV, right ventricle.

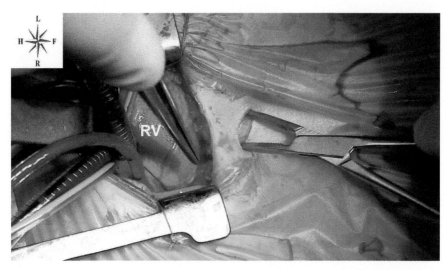

IMAGE 8.44   A small separate incision (0.5 cm, caudally to the main chest incision) is made for the IVC cannula (this incision will be subsequently used for the insertion of the thoracic drainage at the end of the operation). RV, right ventricle.

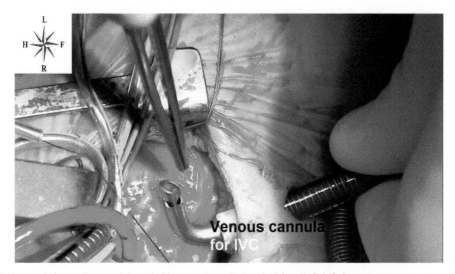

IMAGE 8.45   The IVC cannula is usually passed through this separate small chest incision. IVC, inferior vena cava

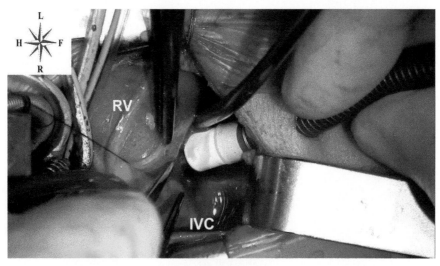

IMAGE 8.46   The IVC cannula is inserted. IVC, inferior vena cava; RV, right ventricle.

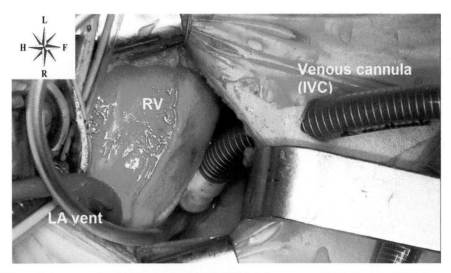

IMAGE 8.47   The IVC cannula is connected and full bypass is achieved. IVC, inferior vena cava; LA, left atrium; RV, right venticle.

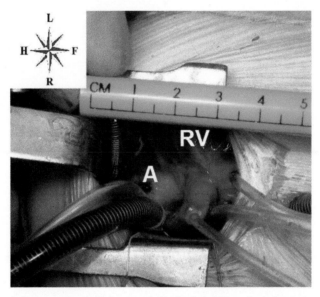

IMAGE 8.48   The Bookwalter retractor is then reconnected offering a good visualization of the mediastinal structures. A, ascending aorta; RV, right ventricle.

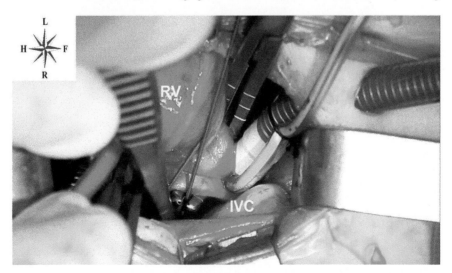

IMAGE 8.49   Vessel loops are placed around the IVC by using a Satinsly clamp. IVC, inferior vena cava; RV, right ventricle.

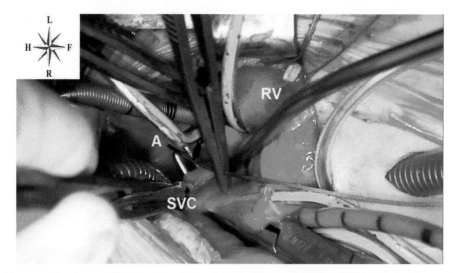

IMAGE 8.50    Vessel loops are placed around SVC by using a right angle clamp. A, ascending aorta; SVC, superior vena cava; RV, right ventricle.

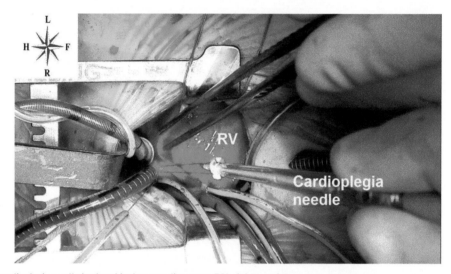

IMAGE 8.51    The cardioplegia needle is placed in the ascending aorta. RV, right ventricle.

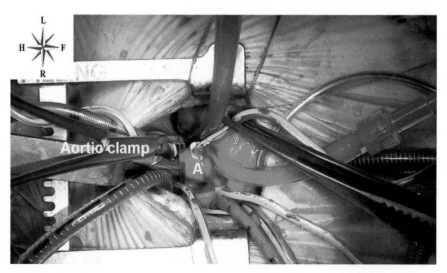

IMAGE 8.52    The aorta is cross clamp is positioned. A, ascending aorta.

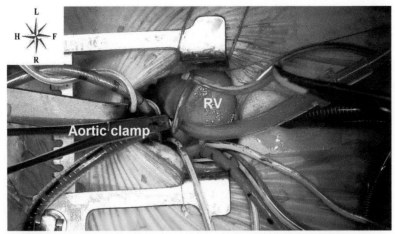

IMAGE 8.53    The aorta is cross clamped and the induction of a cardioplegic arrest. RV, right ventricle.

IMAGE 8.54    Right atriotomy is performed with visualization of the right atrial structure (a pump sucker is visible within the atrial septal communication).

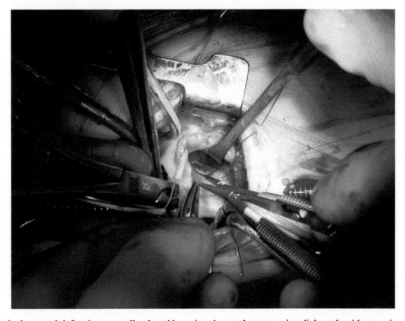

IMAGE 8.55    The ventricular septal defect is eventually closed by using the autologous pericardial patch with a tunning 6.0 polypropylene suture.

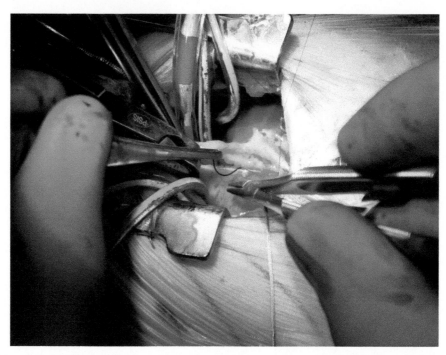

IMAGE 8.56    The right atrial incision in then sutured.

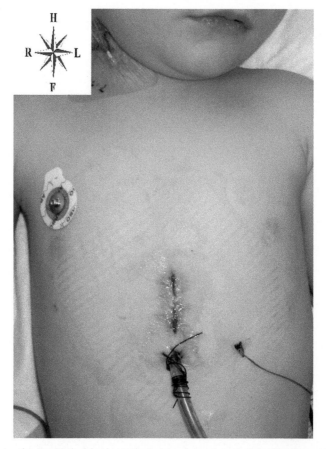

IMAGE 8.57    Postoperative image showing the skin incision in a patient who underwent surgical closure of a ventricular septal defect through a midline lower mini sternotomy.

## Postoperative Cosmetic Result

Postoperative image showing the skin incision in a patient who underwent surgical closure of a ventricular septal defect through a midline lower mini sternotomy (Image 8.57; Video 8.1).

Video 8.1 Midline Lower Mini-Sternotomy (MS).

## REFERENCES

[1] Farhat F, Metton O, Jegaden O. Benefits and complications of total sternotomy and ministernotomy in cardiac surgery. Surg Technol Int. 2004;13:199–205.

[2] Sebastian VA, Guleserian KJ, Leonard SR, Forbess JM. Ministernotomy for repair of congenital cardiac disease. Interact Cardiovasc Thorac Surg 2009;9(5):819–21. Nov.

[3] Vida VL, Padalino MA, Boccuzzo G, Veshti A, Speggiorin S, Falasco G, et al. MD Minimally invasive surgery for congenital heart disease: a gender differentiated approach. J Thorac Cardiovasc Surg 2009;138(4):933–6.

[4] Vida VL, Padalino MA, Motta R, Stellin G. Minimally invasive surgical options in pediatric heart surgery. Expert Rev Cardiovasc Ther 2011;9(6):763–9.

# Chapter 9

# Right Anterior Mini-Thoracotomy (RAMT)

Vladimiro L. Vida, Giovanni Stellin
*University of Padua, Padua, Italy*

## INTRODUCTION

Midline sternotomy is currently utilized as a standard approach for the correction of simple congenital cardiac defects in many centers. Although there are good functional patient's outcomes, this approach often yields poor cosmetic results, which arouses displeasure and psychological distress, especially among young females [1–3].

The use of right anterolateral thoracotomy (Image 9.1) has been advocated since many years, and it has been employed as an alternative to mini-sternotomy [3–11], particularly in the female gender. Although in adult female patients the superiority of this approach is clear, few data are currently present in the medical literature regarding late results of this surgical approach, especially when performed in the prepubescent age. We utilized this technique in children and adults with congenital heart disease mainly for treating ostium secundum type atrial septal defects (ASDs) and partial atrioventricular septal defects, mainly in the females, as an alternative to a MS. Nonetheless, this technique has also been recently employed to treat other more complex congenital heart defects and has been chosen by the majority of males who underwent minimally invasive procedures.

At the beginning of our experience, 6–7-cm semilunar incision was made in the sulcus of the right breast (6–8 cm away from the nipple area), which entered in the chest in the fourth intercostal space. More recently we have minimize the lenght of the surgical incision to 3–4 cm. At prepuberty age, the incision was kept very low under the right nipple, particularly in female patients, with the aim of avoiding any possible future interference with breast development. We strongly suggest that the location of the incision in the submammary area needs to be very low under the right nipple (above the fifth intercostal space), away from any possibility of future breast development.

## SURGICAL TECHNIQUE

### Orientation

**H:** head, **F:** foot, **R:** right, **L:** left.

### Patient Preparation and Right Anterior Mini-Thoracotomy

The patient placed in a supine position with a roll under the right hemithorax. We carefully identified the sternum and the first five intercostal spaces (Image 9.2). After surgical isolation and preparation of femoral vessel for peripheral cannulation, a semilunar incision in the right chest is usually performed. At prepuberty age, the incision is kept very low under the right nipple (usually over the fifth intercostal space), with the aim of avoiding any possible future interference with breast development (Images 9.3 and 9.4). In adolescent females, when the mammary gland is evident, the incision is usually done 5 mm below the mammary sulcus (Images 9.5 and 9.6). Subcutaneous fat and mammary gland are usually gently dissected from the fascia up to the fourth intercostal space, where the chest cavity is entered (in the fourth or fifth intercostal space, depending on the type of congenital heart defect (CHD) to correct, which is usually fifth intercostal space for ASD II and fourth intercostal space for other CHDs in which the aortic clamp may be required). The incision of the intercostal space is about 1 cm longer than the skin incision at each side (Images 9.7–9.9).

Fundamentals of Congenital Minimally Invasive Cardiac Surgery. http://dx.doi.org/10.1016/B978-0-12-811355-4.00009-5
Copyright © 2018 Elsevier Inc. All rights reserved.

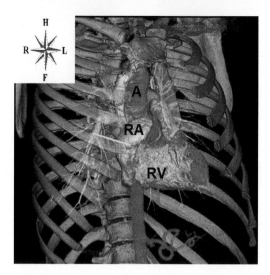

IMAGE 9.1    **CT of the chest with 3D reconstruction showing the relationship between thoracotomy and the right atrial structures.** *Yellow line* shows the right anterior mini-thoracotomy surgical access. A, Aorta; RA, right atrium; RV, right ventricle.

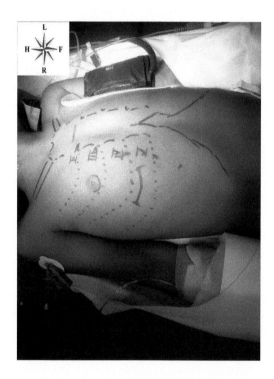

IMAGE 9.2    **The patient placed in a supine position with a roll under the right hemithorax.** We carefully identified the sternum and the first five intercostal spaces.

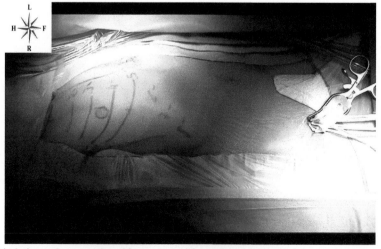

IMAGE 9.3    **Surgical isolation and preparation of femoral vessel for peripheral cannulation.**

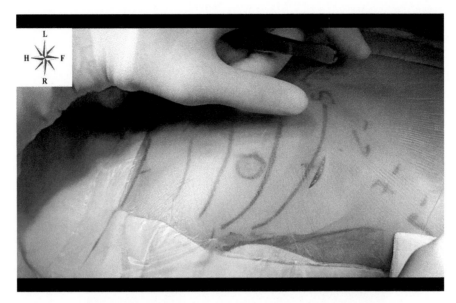

IMAGE 9.4   **A semilunar incision in the right chest is usually performed.** At prepuberty age, the incision is kept very low under the right nipple (usually over the fifth intercostal space), with the aim of avoiding any possible future interference with breast development.

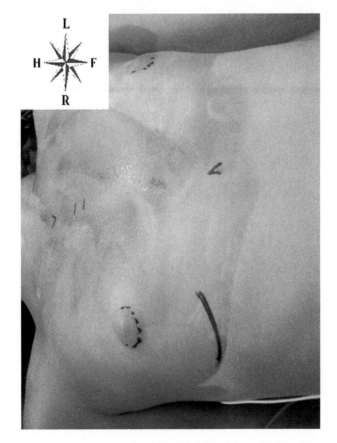

IMAGE 9.5   **In adolescent females, when the mammary gland is evident, the incision is usually done 5 mm below the mammary sulcus.**

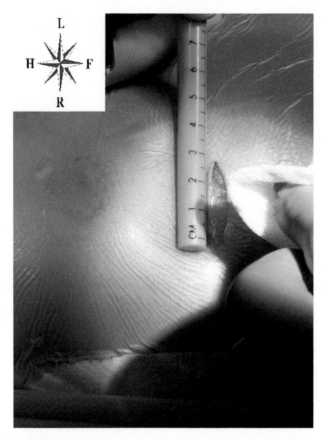

IMAGE 9.6    The incision is usually limited to 2.5–4 cm.

IMAGE 9.7    Subcutaneous fat and mammary gland are usually gently dissected from the fascia up to the fourth intercostal space.

## Chest Retraction

A Finocchietto retractor (Chapter 5) is used to spread the ribs (Image 9.10), and a second retractor (mainly in patients with a body weight less than 20 kg) is positioned orthogonally in order to spread the skin incision (Image 9.11). In patients having more than 20 kg of body weight or in patients with well-represented fat tissue, a soft tissue retractor may be used to enhance surgical vision and to achieve a proper hemostasis of fat and muscular tissue at the incision level (Image 9.12).

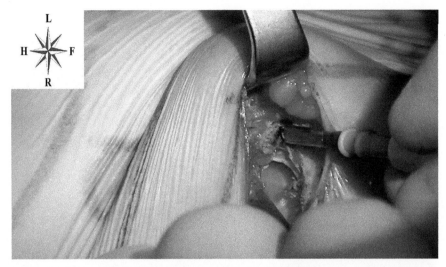

IMAGE 9.8    The chest cavity is entered (in the fourth or fifth intercostal space, depending on the type of congenital heart defect (CHD) to correct, which is usually fifth intercostal space for ASD II and fourth intercostal space for other CHDs in which the aortic clamp may be required).

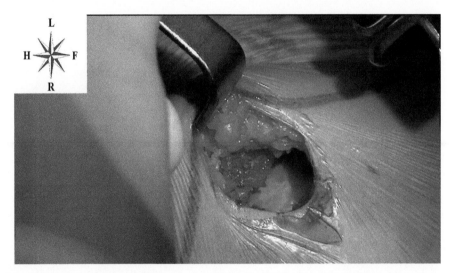

IMAGE 9.9    The incision of the intercostal space is about 1 cm longer than the skin incision at each side.

IMAGE 9.10    A Finocchietto retractor is used to spread the ribs.

IMAGE 9.11   A second retractor (mainly in patients with a body weight less than 20 kg) is positioned orthogonally in order to spread the skin incision.

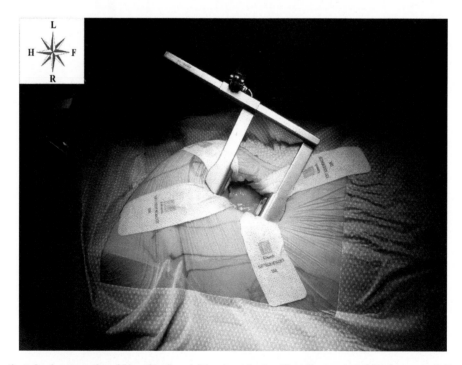

IMAGE 9.12   In patients having more than 20 kg of body weight or in patients with well-represented fat tissue, a soft tissue retractor may be used to enhance surgical vision and to achieve a proper hemostasis of fat and muscular tissue at the incision level.

A large sponge is usually inserted into the right chest to compress temporarily the right lung and enhance the vision of the heart structures (Image 9.13). In bigger patients, right lung exclusion is usually carried on. After cardiopulmonary bypass (CPB) is started and the lung is disconnected, the sponge is removed. *: pericardium.

A proper exposure is also achieved by traction carried out by a "Farabeuf" double-ended retractor attached to the Bookwalter Retractor System (*) (Chapter 4), which is used to lift the ribs. This is particularly important if aortic cannulation is required (Images 9.14 and 9.15).

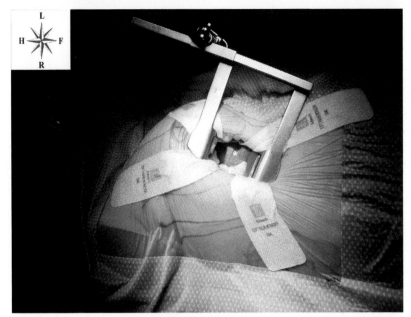

IMAGE 9.13  **A large sponge is usually inserted into the right chest to compress temporarily the right lung and enhance the vision of the heart structures.** In bigger patients, right lung exclusion is usually carried on. *Pericardium

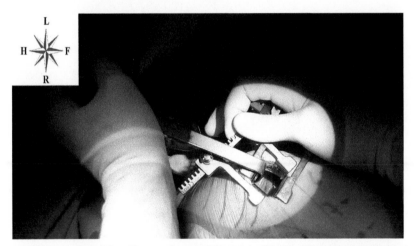

IMAGE 9.14  **A proper exposure is also achieved by traction carried out by a "Farabeuf" double-ended retractor attached to the Bookwalter Retractor System (*) (Chapter 4), which is used to lift the ribs.** This is particularly important if aortic cannulation is required.

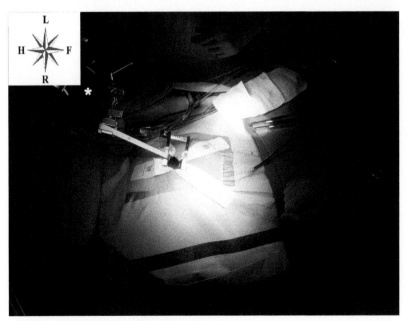

IMAGE 9.15  **Intraoperative view of the Bookwalter Retractor System.**

## Pericardial Patch Harvesting and Suspension

An adequate portion of the pericardium (*) is then harvested and subsequently fixed in glutaraldehyde 6% for 10 min (Image 9.16). The remaining part of the pericardium is then suspended outside the field in order to expand the operative view. Sutures are anchored to the pericardium and carried through the chest wall with special instruments (Spike, Chapter 4) (Images 9.17 and 9.18). Thymectomy is rarely required in right anterior mini-thoracotomy (RAMT).

## Central Aortic Cannulation (When Necessary)

After complete patient heparinization, peripheral cannulation of the femoral artery and veins is usually performed at this stage. Nonetheless, in selected patients (patients with a body weight <15 kg or in patients is whom femoral artery is

IMAGE 9.16   An adequate portion of the pericardium (*) is then harvested and subsequently fixed in glutaraldehyde 6% for 10 min.

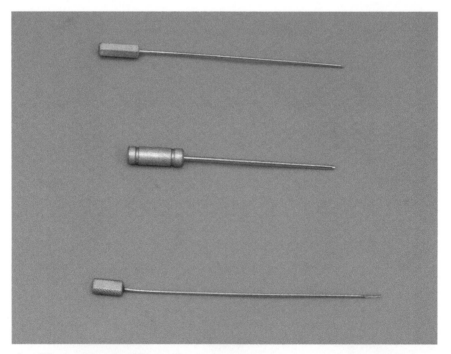

IMAGE 9.17   Spike pericardial retraction system (Chapter 4).

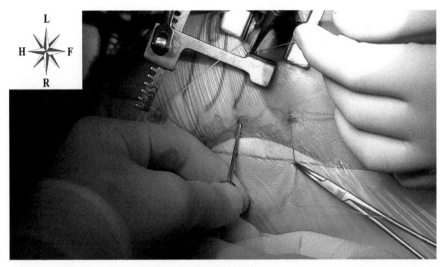

IMAGE 9.18  **The remaining part of the pericardium is then suspended outside the field in order to expand the operative view.** Sutures are anchored to the pericardium and carried through the chest wall with special instruments (Spike pericardium retractor).

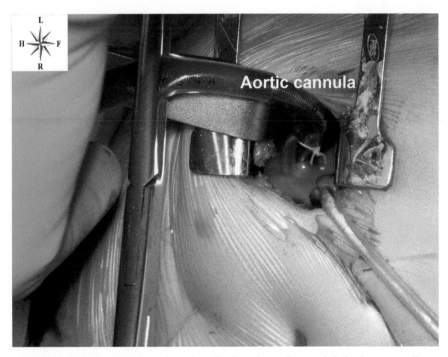

IMAGE 9.19  **In patients with a body weight <15 kg or in patients is whom femoral artery is judged as too small, central aortic cannulation may be required.**

judged as too small), central aortic cannulation may be required. A single longitudinal diamond-shaped tobacco purse-string is fashioned in the ascending aorta, and a Fem-flex cannula is usually utilized (Chapter 4). This cannula has a long stilet, which makes aortic cannulation easy (Image 9.19). After cardiopulmoanary bypass (CPB) is initiated, umbilical tapes are placed around superior vena cava (SVC) and inferior vena cava (IVC). A Satinsky clamp (Chapter 4) is used for encircling the IVC (Images 9.20 and 9.21). A right angle clamp is used to encircle the SVC (Image 9.22). In case the SVC appears to be distant from the surgical incision, a DeBakey clamp, inserted through a separate 5-mm incision (that can be later used to insert the chest tube), can be used to temporarily occlude the SVC (as shown in the fibroscopic view) (Images 9.23 and 9.24).

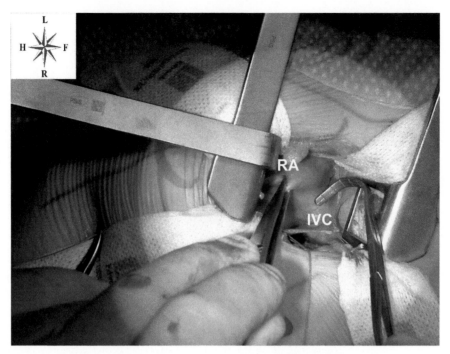

IMAGE 9.20   **A Satinsky clamp is used for encircling the IVC.** *IVC,* Inferior vena cava, *RA,* right atrium.

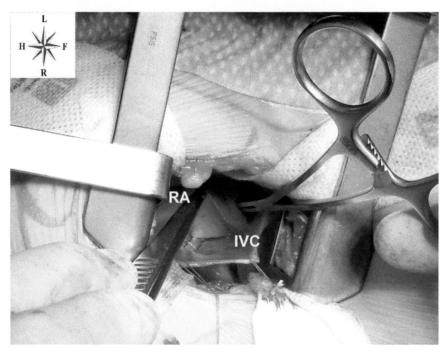

IMAGE 9.21   **The Satinsky clamp is around the IVC.** *IVC,* Inferior vena cava, *RA,* right atrium.

## Induced Ventricular Fibrillation

During RAMT, the preferred method for achieving intracardiac repair (especially in a patient with simple CHD) is to induce a ventricular fibrillation. The plaque (*) for the induction of ventricular fibrillation (Chapter 5) (Images 9.25–9.27) is placed on the diaphragmatic portion of the right ventricle, and ventricular fibrillation (VF) is induced (Images 9.28 and 9.29). The surgical visibility is good even when central cannulation is performed (Image 9.27). During VF, particular care is taken to

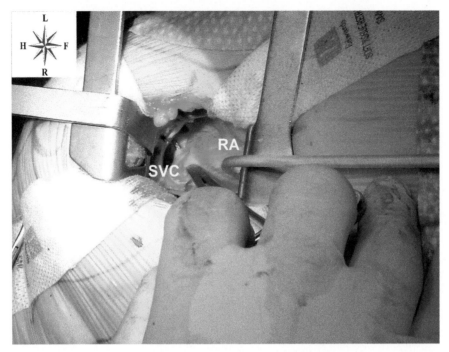

IMAGE 9.22  **A right angle clamp is used to encircle the SVC.** *SVC,* Superior vena cava, *RA,* right atrium.

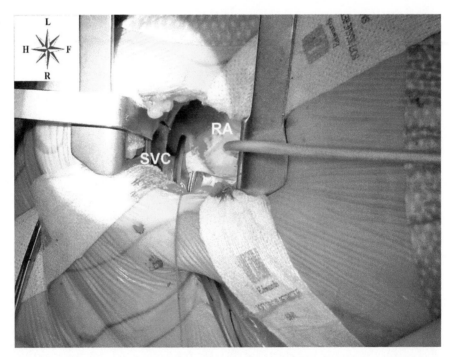

IMAGE 9.23  **In case the SVC appears to be distant from the surgical incision, a DeBakey clamp, inserted through a separate 5-mm incision (that can be later used to insert the chest tube), can be used to temporarily occlude the SVC.** SVC, superior vena cava. *RA,* Right atrium.

avoid pump suctioning into the left atrium. After completion of repair, cardiac rhythm is usually restored spontaneously after short periods of VF, while a lidocaine bolus (1 mg/kg) followed by external DC shocks (3–5 J/kg) is often required in case of VF time prolongation. In case of persistent VF, amiodarone infusion may be indicated (2–4 mg/kg bolus iv.). Particular care is taken regarding the deairing of the left heart chamber, under 2D-TEE echography guidance, before the heart starts to beat. A: aortic cannula. Black cable: fibrillator cable.

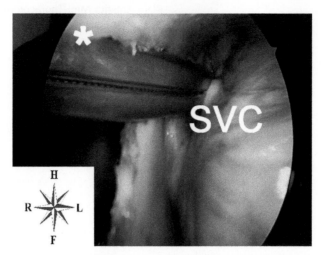

IMAGE 9.24   **Fibroscopis view showing the encircling of the SVC by using a DeBakey clamp (\*).** *SVC*, Superior vena cava.

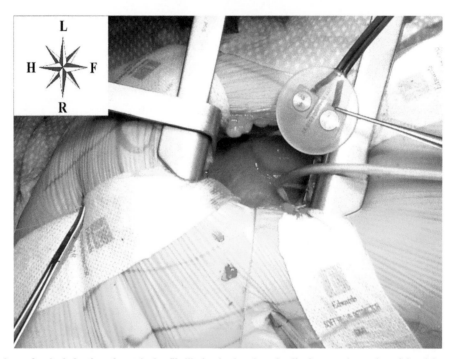

IMAGE 9.25   **The plaque for the induction of ventricular fibrillation is placed on the diaphragmatic portion of the right ventricle.**

IMAGE 9.26   **The plaque (\*) for the induction of ventricular fibrillation is smaller for patients with a body weight below 15 kg.**

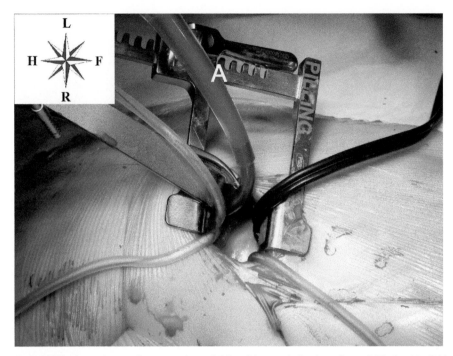

IMAGE 9.27    **The surgical visibility is good even when central arterial (aortic) cannulation is performed.** *Black cable*, Cable for inducing ventricular fibrillation. A: Aortic cannula

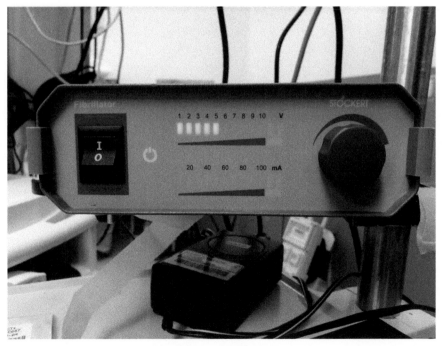

IMAGE 9.28    **Fibrillator Fi 20 M, Stockert (Sorin Group, Munchen) (Chapter 4).**

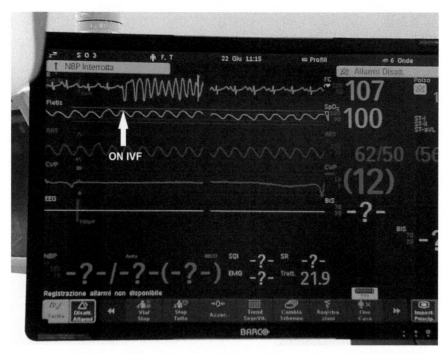

IMAGE 9.29    **Intraoperative monitoring of the cardiac rhythm showin the induction of ventricular fibrillation** (*white arrow*).

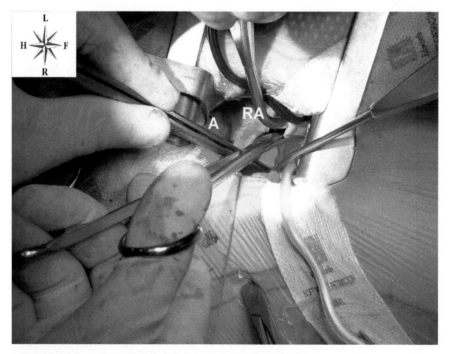

IMAGE 9.30    **Transversal right atriotomy is performed.** *A*, Ascending aorta; *RA*, right atrium.

## Atriotomy and Intracardiac Repair

Transversal right atriotomy is performed, and the atrial wall is suspended with separate stitches at the extremities (Images 9.30 and 9.31). The atrial septal defect is identified (Image 9.32) and subsequently closed. A, indicates aorta and RA indicates right atrium. White arrow indicates the atrial septal defect. Care in taken to properly deair the left atrial structures (under 2D TEE guidance) before complete septation.VIDEO Right-anterior mini-thoracotomy (1).wmv

IMAGE 9.31   **The right atrial wall is suspended with separate stitches at the extremities.** *RA*, Right atrium.

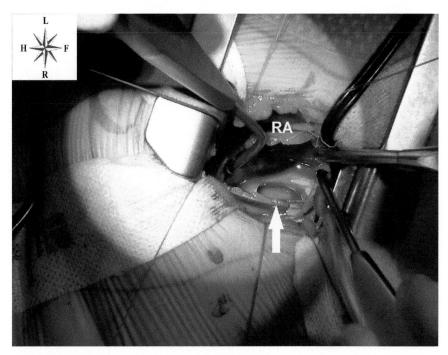

IMAGE 9.32   **The atrial septal defect (*white arrow*) is identified and subsequently closed.** *RA*: Right atrium.

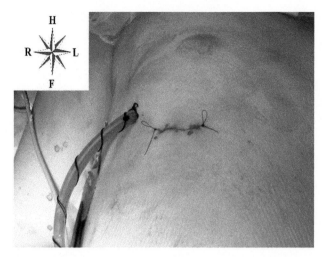

IMAGE 9.33   **Final status of the skin incision at the end of surgical procedure.**

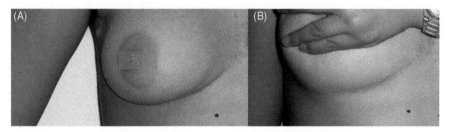

IMAGE 9.34   **Postoperative image of the RAMT incision.** At the beginning of our experience the surgical incision was longer (about 6–8 cm) and more prolonged toward the midline of the chest.

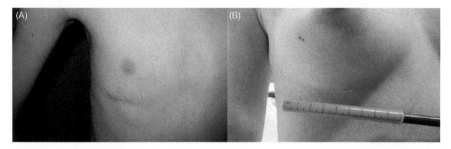

IMAGE 9.35   **Postoperative image of the RAMT incision.** In more recent experiences, we have minimized the size of the surgical incision (usually 3–4 cm), which was more lateral.

## Postoperative Cosmetic Result

Final status of the skin incision at the end of surgical procedure in shown in Image 9.33.

Postoperative image shows the evolution of the RAMT incision. At the beginning of our experience the surgical incision was longer (about 6–8 cm) and more prolonged toward the midline of the chest (Image 9.34). In more recent experiences, we have minimized the size of the surgical incision (usually 3–4 cm), which was more lateral (Image 9.35).

## REFERENCES

[1]  Hagl C, Stock U, Haverich A, Steinhoff G. Evaluation of different minimally invasive techniques in pediatric cardiac surgery. Is a full sternotomy always a necessity? Chest 2001;119(2):622–7.

[2]  Lancaster LL, Mavroudis C, Rees AH, Slater AD, Ganzel BL, Gray LA Jr. Surgical approach to atrial septal defect in the female. Right thoracotomy versus sternotomy. Am Surg 1990;56:218–21.

[3]  Vida VL, Padalino MA, Boccuzzo G, Veshti AA, Speggiorin S, Falasco G, et al. Minimally invasive operaton for congenital heart disease: a sex-differentiated approach. J Thorac Cardiovasc Surg 2009;933–6.

[4]  Vida VL, Padalino MA, Motta R, Stellin G. Minimally invasive surgical options in pediatric heart surgery. Expert Rev Cardiovasc Ther 2011;9(6):763–9.

[5]  Metras D, Kreitmann B. Correction of cardiac defects through a right thoracotomy in children. J Thorac Cardiovasc Surg 1999;117(5):1040–2.

[6]  Lancaster LL, Mavroudis C, Rees AH, Slater AD, Ganzel BL, Gray LA Jr. Surgical approach to atrial septal defect in the female. Right thoracotomy versus sternotomy. Am Surg 1990;56(4):218–21.

[7]  Abdel-Rahman U, Wimmer-Greinecker G, Matheis G, Klesius A, Seitz U, Hofstetter R, et al. Correction of simple congenital heart defects in infants and children through a minithoracotomy. Ann Thorac Surg 2001;72(5):1645–9.

[8]  Dabritz S, Sachweh J, Walter M, Messmer BJ. Closure of atrial septal defects via a limited right anterolateral thoracotomy as a minimal invasive approach in female patients. Eur J Cardiothorac Surg 1999;15:18–23.

[9]  Mishaly D, Ghosh P, Preisman S. Minimally invasive congenital cardiac surgery through right anterior minithoracotomy approach. Ann Thorac Surg 2008;85:831–5.

[10]  Cheng DC, Martin J, Lal A, Diegeler A, Folliguet TA, Nifong LW, et al. Minimally invasive versus conventional open mitral valve surgery: a meta-analysis and systematic review. Innovations (Phila) 2011;6(2):84–103.

[11]  Vida VL, Tessari C, Fabozzo A, Padalino MA, Barzon E, Zucchetta F, et al. The evolution of the right anterolateral thoracotomy technique for correction of atrial septal defects: cosmetic and functional results in prepubescent patients. Ann Thorac Surg 2013;95(1):242–7.

Chapter 10

# Right Lateral Mini-Thoracotomy (RLMT)

Vladimiro L. Vida, Alvise Guariento, Giovanni Stellin
*University of Padua, Padua, Italy*

## SURGICAL TECHNIQUE

### Orientation

**H:** head, **F:** foot, **R:** right, **L:** left.

Since 2007, in addition to a lower mid-line mini-sternotomy and a right anterior mini-thoracotomy, a right posterior-lateral mini-thoracotomy has also been offered as a surgical option (1–5). This technique has been described by Metras et al. in 1999 (6) for repairing simple congenital heart disease (CHD) and has been modified by us from a classic wide right posterior thoracotomy to a minimally invasive approach. Initially, a 4 cm subscapular incision starting from the angle of the scapula and extending anteriorly was employed for entering the chest in the fourth intercostal space (Image 10.1). Over the time, we moved the incision more laterally, in subaxillary position, between the anterior and the median axillary lines with the lower end at the level of the mammillary line (Image 10.2). Depending on the type of CHD, we may enter into the chest by using the fourth or the fifth intercostal space. In details, we enter the chest in the fifth intercostal space for treating more simple CHDs as ASD II and partial-AVSD, usually with the use of induced ventricular fibrillation. In contract, we chose to enter in the fourth intercostal space for more complex CHDs as partial anomalous pulmonary venous connection (PAPVC), discrete subaortic stenosis, ventricular septal defects, and so on. Through this approach the aorta can be easily visualized and cross-clamped. As in right anterior mini-thoracotomy (RAMT), when aortic cannulation is required (patients with a body weight <15 kg or in patients where femoral artery is judged too small) a single longitudinal diamond-shape tobacco purse-string is performed in the ascending aorta. A Fem-flex cannula is usually utilized (Chapter 4). This cannula has a long stilet, which make easy aortic cannulation.

### Patient Preparation and RLMT

The patient is prepare in supine position. The site of the incision is marked; we identify the anterior and the mid axillary lines. The surgical incision will follow the direction of the ribs. It is important to mark the chest in this position because the reationship of the soft tissue with the chest cage may chage after positioning the patient for surgery in left lateral decubitus (Images 10.3 and 10.4). After surgical isolation and preparation of femoral vessel for peripheral cannulation, a 3–4 cm incision is performed in the right chest in the right chest. Subcutaneous fat and mammary gland are usually gently dissected from the fascia up to the fourth intercostal space, where the chest cavity is entered (just above the right atrial structures). The incision of the intercostal space is about 1 cm longer of the skin incision at each side (Images 10.5–10.9). Video 10.1 Right lateral (axillary) mini-thoracotomy approach in a patient with partial anomalous pulmonary venous connections.

According to the type of CHD, we may enter into the chest by using the fourth or the fifth intercostal space. We usually choose to enter the chest in the fifth intercostal space for treating more simple CHDs as ASD II and partial atrioventricular septal defects, usually opening the heart by using an induced ventricular fibrillation. The choice of entering the chest in the fourth intercostal space is reserve for treating for more complex CHHs as PAPVC, discrete subaortic stenosis, ventricular septal defects, and so on. where aortic cross clamping followed by cardioplegic arrest is required.

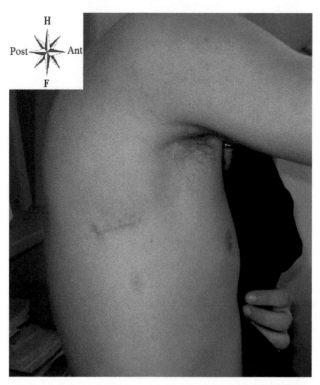

IMAGE 10.1   **Postoperative image: 16 year-old patients who underwent correction of a partial anomalous pulmonary venous connection via a right lateral mini-thoracotomy (RLT).**

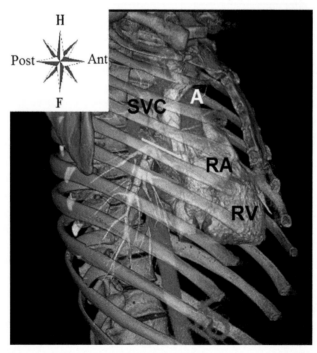

IMAGE 10.2   **CT of the chest with 3D reconstruction showing the relationship between the thoracotomy and the cardiac structures.** *Yellow* and *green line* showing the RLT surgical access, fourth and fifth intercostal spaces respectively. *A,* Aorta; *RA,* right atrium; *RV,* right ventricle; *SVC,* superior vena cava.

A Finocchietto retractor (Chapter 4) is used to spread the ribs (Image 10.10) and a second retractor is positioned orthogonally (Image 10.11) to spread the skin incision. A soft tissue retractor can be employed in older patients (with more fat tissue); it enhance a better visualization of intrathoracic structures and also contributed to haemostasis of the subcutaneous and muscular tissue. Lung exclusion is usually carried on in patients with a body weight above 20 kg.

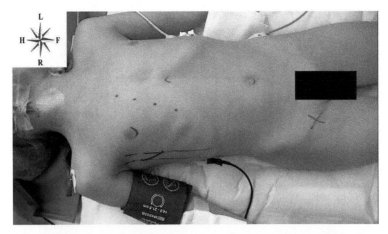

IMAGE 10.3  **Operative position of a patient who will undergo a RLMT.** After initial anesthesiology preparation and skin marking of the surgical sites.

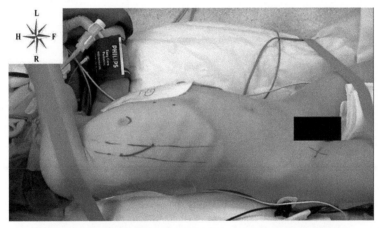

IMAGE 10.4  **Operative position of a patient who will undergo a RLMT.** The patients is rotated in left lateral decubitus.

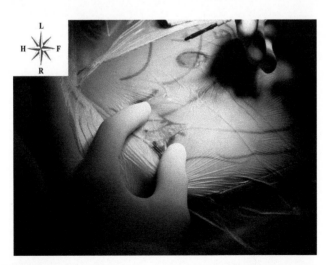

IMAGE 10.5  **After surgical isolation and preparation of femoral vessel for peripheral cannulation, a 3–4 cm incision is performed in the right chest in the right chest.**

After a portion of the pericardium is then harvested and subsequently fixed in glutaradehyde 6% for 10 min. The in situ pericardium is suspended outside the field to expand the operative view. Sutured are anchored to the pericardium and carried through the chest wall with a special instruments (Spike, see Chapter 4) (Images 10.12–10.15). Thymectomy is rarely required in right lateral mini-thoracotomy (RLMT).

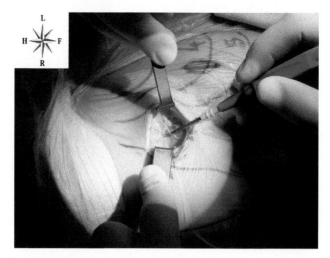

IMAGE 10.6   **Subcutaneous fat and mammary gland are usually gently dissected.**

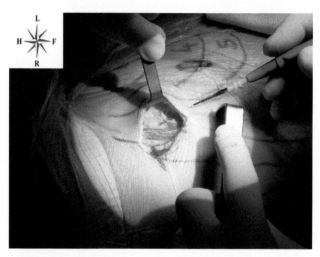

IMAGE 10.7   **The muscles are usually spared during dissection.**

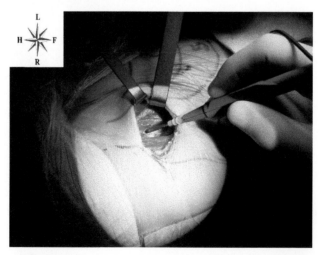

IMAGE 10.8   **The intercostal spece is identified.** We chose to enter the chest in the fourth or fifth intercostal space, depending on the type of congenital heart defect to correct. It is usually the fifth intercostal space for the corection of atrial septal defects and the fourth intercostal space for other congenial heart diseases in which the aortic clamp may be required.

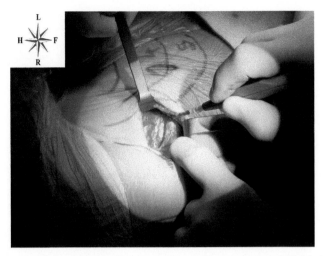

IMAGE 10.9   **The chest cavity is entered.** The incision of the intercostal space is about 1 cm longer of the skin incision at each side.

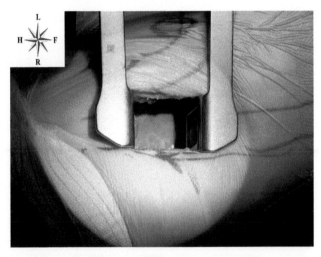

IMAGE 10.10   **A Finocchietto retractor (Chapter 4) is used to spread the ribs.**

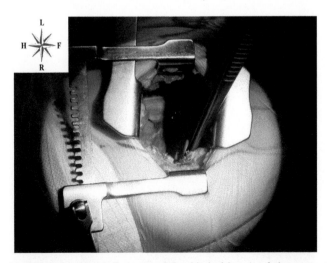

IMAGE 10.11   **A second retractor is positioned orthogonally to spread the skin incision.** A soft tissue retractor can be employed in older patients (with more fat tissue); it enhance a better visualization of intrathoracic structures and also contributed to haemostasis of the subcutaneous and muscular tissue. Lung exclusion is usually carried on in patients with a body weight above 20 kg.

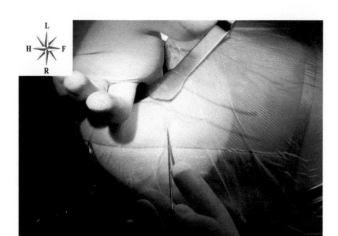

IMAGE 10.12

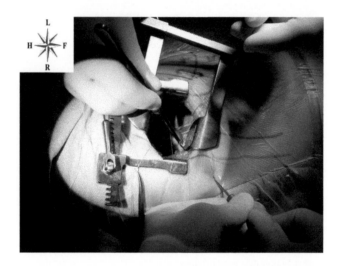

IMAGE 10.13

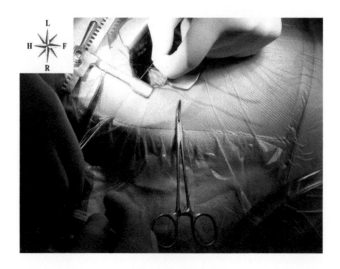

IMAGE 10.14

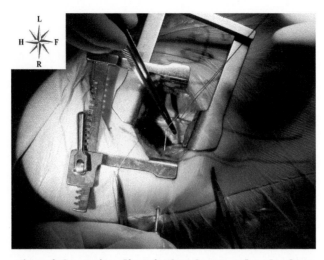

IMAGES 10.12-10.15  **After a portion of the pericardium is then harvested and subsequently fixed in glutaradehyde 6% for 10 min.** The in situ pericardium is suspended outside the field to expand the operative view (see Chapter 4). Sutured are anchored to the pericardium and carried through the chest wall with a special instruments (Spike, see Chapter 4).

After complete patient heparinization peripheral cannulation of femoral artery and venis usually performed at this stage. Nonethless, in selected patients (patients with a body weight <15 kg or in patients where femoral artery is judged too small) central aortic cannulation may be required. A single longitudinal diamond-shape tobacco purse-string is performed in the ascending aorta and a Fem-flex cannula is usually utilized (Chapter 4). This cannula has a long stilet, which make easy aortic cannulation. After cardio-pulmonary bypass (CPB) is initated, umbilical tapes (*) are placed around superior vena cava (SVC) and inferior vena cava (IVC). A Satinsly clamp (Chapter 4) is used for encircling the IVC (fibroscopic images) (Images 10.16–10.17).

A right angle clamp is used to encircle the SVC. Although sylastic loop encircling the phrenic nerve (Image 10.18).

The ascending aorta (A) is usually encircled with an umbilical tape (*) and a tobacco purse string (ps) is done for the cardioplegia needle (Images 10.18–10.20) SVC: superior vena cava, RA: right atrium.

The aortic cross clamp (*) can be done with the Cygnet aortic cross clamp (Novare Surgical System, United States) which is directly inserted through the surgical access or (more frequently in patients with a body weight less than 30 kg) by means of a DeBakey clamp (especially in smaller patients) which is inserted through a separate 3 mm incision within the axillary cave (Image 10.20). The cardioplegic solution is than delivered by a 60 cm long cardioplegia needle is employed (Maquet Cardioplegia Needle, Germany). A: ascending aorta.

## Right Atriotomy and Intracardiac Correction

Caval tapes are snared and the RA is opened through a transversal incision. A: aorta, *: aortic cross clamp Image 10.21.

The sinus septal defect is identified (*) (lower part of the fibroscopic image) together with the anomalous drainage of the upper and middle pulmonary veins of the right lung (1 and 2). In the upper-right portion of the image the tricuspid valve (Image 10.22). The partial anomalous pulmonary venous drainage is baffled into the left atrium through the atrial communication with a pericadial patch (p). The patent foramen ovale is closed separately (*) (Images 10.23–10.25). The right atriotomy is closed with a double suture, caval tape are unsnared and the aortic clamp is then removed, after carefully 2D TEE guided deairing of the left heart chamber (Images 10.26 and 10.27).

## Postoperative Cosmetic Result

Surgical incision at the end of surgical procedure (Image 10.28) and one-month postoperative image showing the cosmetic result (Image 10.29).

IMAGE 10.16 **After cardio-pulmonary bypass is initiated an umbilical tape (\*) is placed around the inferior vena cava (IVC).** A Satinsly clamp (Chapter 5) is used for encircling the IVC (fibroscopic images). *IVC*, Inferior vena cava; *RA*, right atrium.

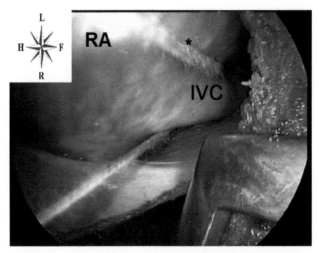

IMAGE 10.17 **A Satinsly clamp (Chapter 4) is used for encircling the IVC (fibroscopic images).**

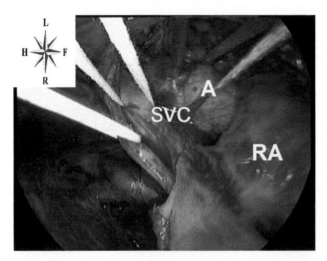

IMAGE 10.18 **A right angle clamp is then used to encircle the superior vena cava.** A white sylastic loop is encircling the phrenic nerve. An umbilical tape is surrounding the ascending aorta. *A*, Ascending aorta; *RA*, right atrium; *SVC*, superior vena cava.

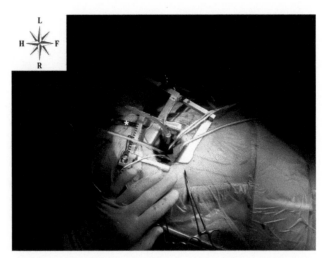

**IMAGE 10.19    Operative view showing the umbilical tapes used for encircling both venae cavae and the ascending aorta.** The cardioplegia needle has been positioned into the ascending aorta. *A*, Ascending aorta; *RA*, right atrium; *SVC*, superior vena cava.

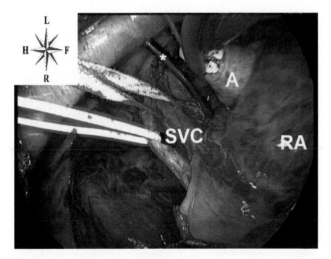

**IMAGE 10.20    The aortic cross clamp (\*) is inserted through a separate 3 mm incision within the axillary cave.** The cardioplega needle is in site in the ascending aorta. *A*, Ascending aorta; *RA*, right atrium; *SVC,* superior vena cava.

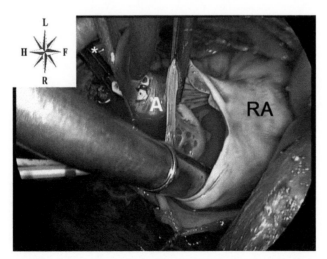

**IMAGE 10.21    Caval tapes are snared and the right atrium is opened through a transversal incision.** *A*, Aorta; *RA*, right atrium; \*, aortic cross clamp.

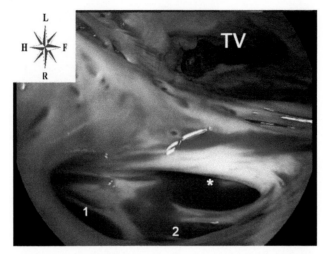

IMAGE 10.22    **The sinus venosus septal defect is identified (\*) (fibroscopic image) together with the anomalous drainage of the upper and middle pulmonary veins of the right lung (1 and 2).** In the upper-right portion of the image the tricuspid valve. *TV*, Tricuspid valve.

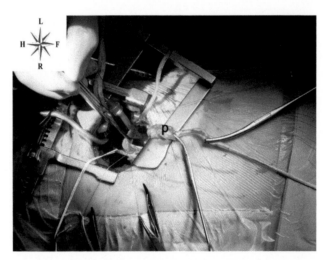

IMAGE 10.23    **The partial anomalous pulmonary venous drainage is baffled into the left atrium through the atrial communication with a pericadial patch (p).**

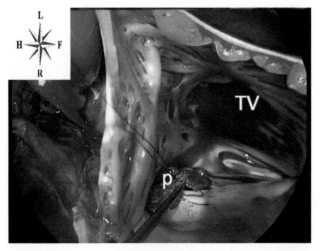

IMAGE 10.24    **Fibroscopic image showing the last phase of baffling of the anomalous pulmonary veins into the left atrium.** *p*, Pericardial patch; *TV*, tricuspid valve.

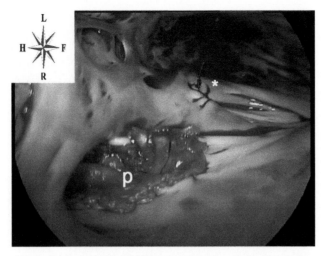

IMAGE 10.25    **The patent foramen ovale is closed separately (\*)(fibroscopic image).** *p*, Pericardial patch.

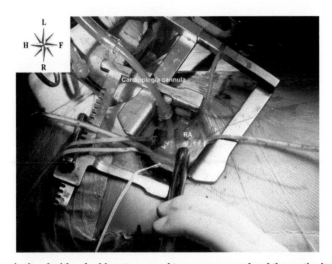

IMAGE 10.26    **The right atriotomy is closed with a double suture, caval tape are unsnared and the aortic clamp is then removed, after carefully 2D TEE guided deairing of the left heart chamber.** A vent is placed into the ascending aorta for deairing. *RA*: Right atrium.

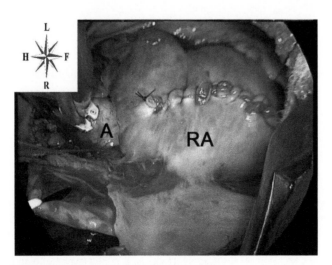

IMAGE 10.27    **Fibroscopic image showing the right atrial incision.** *RA*, Right atrium.

IMAGE 10.28 **Postoperative image showing the length of the surgical incision.**

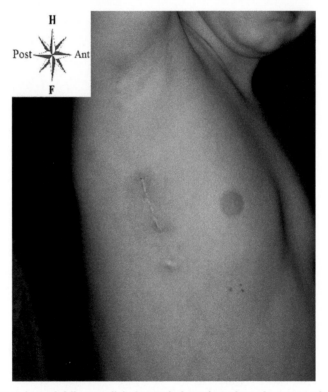

IMAGE 10.29 **Postoperative image at 1 month follow-up clinical control of a patient who underwent patch closure of an atrial septal communication via a RALT.**

## FURTHER READING

[1] Vida VL, Padalino MA, Motta R, Stellin G. Minimally invasive surgical options in pediatric heart surgery. Expert Rev Cardiovasc Ther 2011;9(6):763–9.

[2] Vida VL, Padalino MA, Bhattarai A, Stellin G. Right posterior-lateral mini-thoracotomy access for treating congenital heart disease. Ann Thorac Surg 2011;92(6):2278–80.

[3] Schreiber C, Bleiziffer S, Kostolny M, Hörer J, Eicken A, Holper K, Tassani-Prell P, Lange R. Minimally invasive midaxillary muscle sparing thoracotomy for atrial septal defect closure in prepubescent patients. Ann Thorac Surg 2005;80(2):673–6.

[4] Yang X, Wang D, Wu Q. Repair of atrial septal defect through a minimal right vertical infraaxillary thoracotomy in a beating heart. Ann Thorac Surg 2001;71:2053–4.

[5] Umakanthan R, Petracek MR, Leacche M, Solenkova NV, Eagle SS, Thompson A, Ahmad RM, Greelish JP, Ball SK, Hoff SJ, Absi TS, Balaguer JM, Byrne JG. Minimally invasive right lateral thoracotomy without aortic cross-clamping: an attractive alternative to repeat sternotomy for reoperative mitral valve surgery. J Heart Valve Dis 2010;19(2):236–43.

[6] Metras D, Kreitmann B. Correction of cardiac defects through a right thoracotomy in children. J Thorac Cardiovasc Surg 1999;117(5).

# Chapter 11

# Other Less Commonly Used Minimally Invasive Surgical Approaches

## (A) Upper Mini-Sternotomy

Juan M. Gil-Jaurena, Ana Pita-Fernández, Maria T. González-López, Ramon Pérez-Caballero
*Hospital Gregorio Marañón, Madrid, Spain*

## SURGICAL TECHNIQUE

### Orientation

**H:** head, **F:** foot, **R:** right, **L:** left

The skin incision starts below the angle of Lewis and ends in the mid-point between both nipples (2nd–4th intercostal spaces) ranging 4–5 cm in length on average. The sternum is opened from the sternal notch downward in a "J" or inverted "T" fashion, for which the oscillating saw is recommended, until the 3rd or 4th intercostal space. The thymus is removed at surgeon discretion and the pericardium opened longitudinally. Blunt dissection under the sternum is carried out and a subxyphoideal drain is placed for $CO_2$ delivery throughout the procedure (and surgical drain, eventually). The pericardium is tackled to both sides of the sternum and a craddle is created. Particularly in young infants, aorta and right atrial appendage are easily available at this stage. Purse-strings in ascending aorta and right appendage are performed to cannulate and begin the cardio-pulmonary bypass. We find very useful to cannulate the pulmonary trunk with a left atrial vent catheter, instead of diverting blood from the right superior pulmonary vein. A cardioplegia line and the clamp in the ascending aorta fulfill the surgical scenario. Once the heart is completely arrested, switching the cardioplegia line to a suction catheter can help in emptying the left chambers. The aorta is opened in a standard oblique fashion, the valve is explored meticulously and the procedure customly carried out. On finishing the repair, the aorta is closed in layers, the left chambers deaired as usual (with the aid of $CO_2$ and TEE echocardiography) and cardio-pulmonary bypass discontinued. Modified ultrafiltration (MUF) is a routine in our practice.

Pacing-wires are better implanted before coming-off bypass. The former $CO_2$ delivery line turns out to be the surgical drain. Pericardium is closed and wires are applied for sternum reconstruction, the lower one in a "U" fashion so as to gather the three bone pieces altogether.

On the whole, "simple" aortic cases for which only two/three spots are needed for cannulation (ascending aorta, right atrial appendage, and pulmonary trunk) are best candidates for a upper mini-sternotomy approach. It does not add length to the procedure and the recovery in the Intensive Care Unit and the Ward are optimal (Images 11.1–11.15).

Fundamentals of Congenital Minimally Invasive Cardiac Surgery. http://dx.doi.org/10.1016/B978-0-12-811355-4.00021-6
Copyright © 2018 Elsevier Inc. All rights reserved.

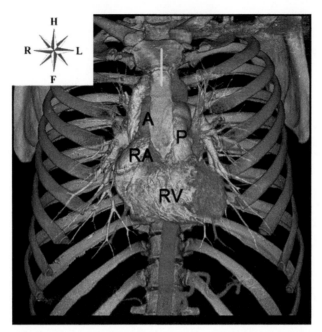

IMAGE 11.1    **Angio-CT scan 3D reconstruction in a control patient showing the relationship between the sternum and the anterior mediastinal structures.** *Yellow-line* showing the position of the mid-line skin incision. *A*, Aorta; *P*, main pulmonary artery trunk; *RA*, right atrium; *RV*, right ventricle.

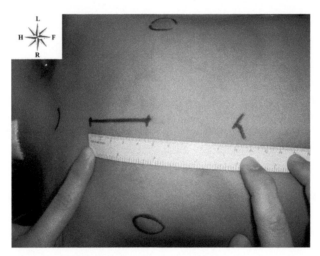

IMAGE 11.2    **Proposed skin incision, from Lewis  angle to mid-point between nipples.**

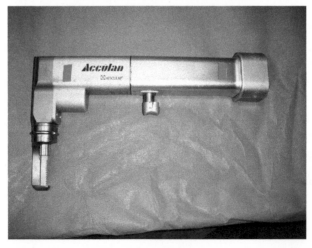

IMAGE 11.3    **Sternal saw driven upside down, from supraclavicular notch downward to 3rd/4th intercostal space level.**

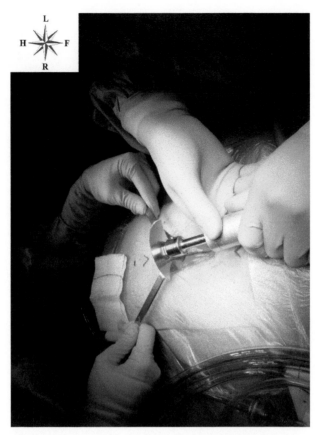

IMAGE 11.4   **Oscillating saw for the horizontal plane in the inverted "T" sternal opening (3rd/4th intercostal space level).**

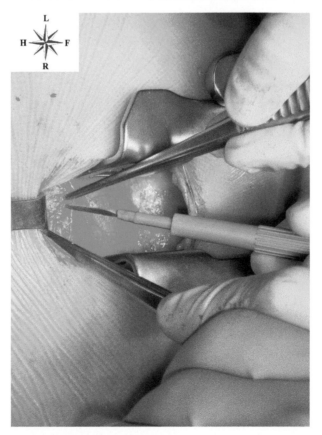

IMAGE 11.5   **The thymus gland (T) is usually partially removed.**

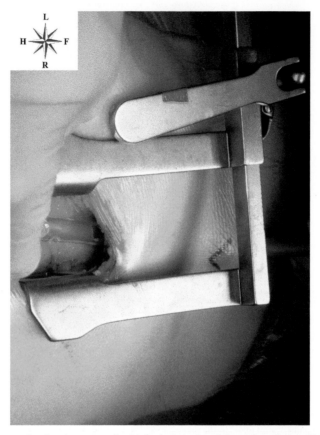

IMAGE 11.6 **After blunt dissection under the sternum, a chest tube in introduced from the subxiphoid for $CO_2$ delivery.** Eventually it will become the chest drainage.

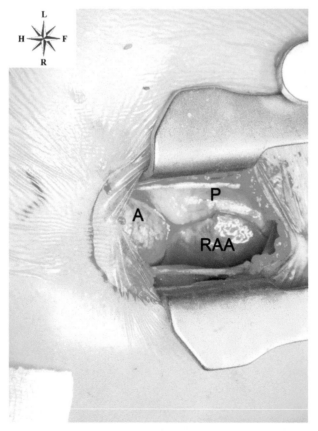

IMAGE 11.7 **Pericardial cradle; displaying ascending aorta (A), right atrial appendage (RAA), and pulmonary artery with infundibulum (P).** All of them easily available.

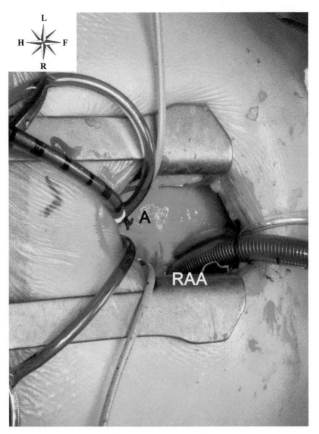

**IMAGE 11.8    Ascending aorta (AA) and right appendage (RAA) are annulated in a straightforward fashion.** As an alternative a peripheral cannulation of the femoral vessels may be proposed.

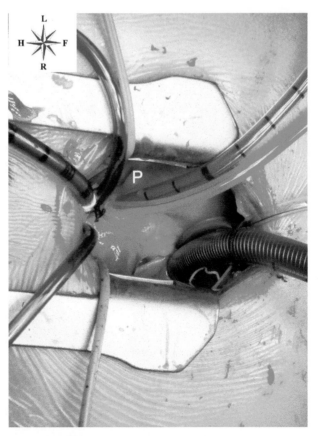

**IMAGE 11.9    Vent catheter in pulmonary artery (P).**

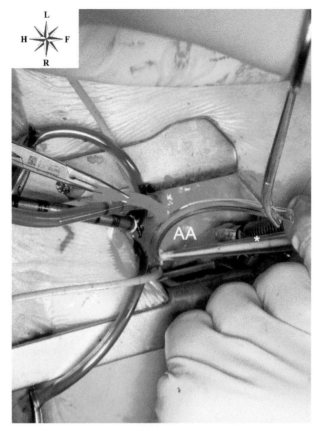

IMAGE 11.10    **Cardioplegia needle (\*) is positioned into the ascending aorta (AA).**

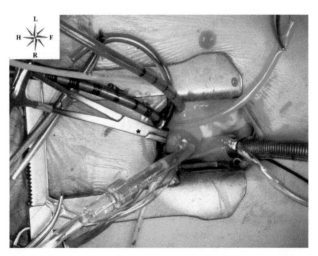

IMAGE 11.11    **The aorta is cross-clamped (\*) and a cold cardioplegic arrest is induced.**

IMAGE 11.12    **Transverse aortotomy is done.** Cardioplegia can be delivered selectively into the coronary ostia in case of significant aortic valve. *: cannula for selective coronary artery cardioplegia delivery.

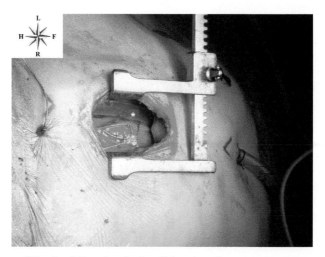

IMAGE 11.13    **Chest drain (*) (former $CO_2$ pipe delivery) and epicardial pacing-wire.**

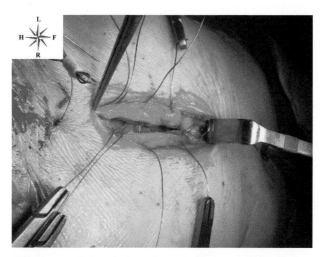

IMAGE 11.14    **Chest closure, with a "U" shape caudad steel wire to fasten the inverted "T" sternum.**

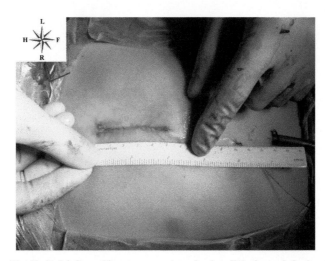

**IMAGE 11.15** Final appearance with a limited 4–5 cm skin scar, suprasternal epicardial wire and single subxiphoideal chest drainage.

## REFERENCES

[1] Mihaljevic T, Cohn LH, Unic D, Aranki SF, Couper GS, Byrne JG. One thousand minimally invasive valve operations: early and late results. Ann Surg 2004;240:529–34.

[2] Tabata M, Umakanthan R, Cohn LH, Bolman RM III, Shekar PS, Chen FY, et al. Early and late outcomes of 1000 minimally invasive aortic valve operations. Eur J Cardiothorac Surg 2008;33:537–41.

[3] Murtuza B, Pepper JR, Stanbridge RD, Jones C, Rao C, Darzi A, et al. Minimal access aortic valve replacement: is it worth it? Ann Thorac Surg 2008;85:1121–31.

[4] Schmitto JD, Mokashi SA, Cohn LH. Minimally-invasive valve surgery. J Am Coll Cardiol 2010;56(6):455–62.

[5] Perrotta S, Lentini S. Ministernotomy approach for surgery of the aortic root and ascending aorta. Interact Cardiovasc Thorac Surg 2009;9(5):849–58.

[6] Karimov JH, Santarelli F, Murzi M, Glauber M. A technique of an upper V-type ministernotomy in the second intercostal space. Interact Cardiovasc Thorac Surg 2009;9(6):1021–2.

[7] Gil-Jaurena JM, González-López MT, Pérez-Caballero R, Pita A, Castillo R, Miró L. 15 years of minimally invasive paediatric cardiac surgery; development and trends. An Pediatr 2016;84(6):304–10.

[8] Gil-Jaurena JM, Pérez-Caballero R, Pita-Fernández A, González-López MT, Sánchez J, De Agustín JC. How to set-up a program of minimally-invasive surgery for congenital heart defects. Transl Pediatr 2016;5(3):125–33.

Chapter 12

# Surgical Results: A Single-Centre 20-year Experience

Vladimiro L. Vida, Chiara Tessari, Massimo A. Padalino, Giovanni Stellin
*University of Padua, Padua, Italy*

From AUGUST 1996 to DECEMBER 2017, we were able to treat 910 patients affected by congenital heart defects (CHDs) with minimally invasive surgical techniques. There were 550 females (60%) and 360 males (40%). Median age at repair was 5 years [median and interquartile range (IQR) 1.7–11 years]. Main diagnosis leading of surgery was an atrial septal defect ostium secundum type (ASD II) in 543 patients (60%), a ventricular septal defect (VSD) in 190 patients (21%), a partial atrio-ventricular septal defect (p-AVSD) in 70 patients (7.7%), a partial anomalous pulmonary venous connection (PAPVC) in 58 patients (6.4%) (Fig. 12.1). Other less common CHDs (4.9%), which were treated by minimally invasive surgical approaches. Minimally invasive access was a right anterior mini-thoracotomy (RAMT) in 427 patients (47%) (Table 12.1), a mid-line MS in 349 patients (38%) (Table 12.2) and a right-lateral (axillary) mini-thoracotomy in 134 patients (15%) (in our more recent experience) (Table 12.3).

All patients required the use of CPB. In 564 (62%), we used an induced ventricular fibrillation (IVF) to achieve intra-cardiac repair while in 344 patients (38%) the aorta was clamped and a cardioplegic arrest was induced.

There was no early and late mortality. Seventy-five patients had postoperative complications (8.2%) (Table 12.4). Median intensive care unit (ICU) time was 1 day (IQR 1–2 days) and median hospitalization time was 6 days (IQR 5–7).

## MID-LINE MINI-STERNOTOMY

A MS was performed in 427 patients (47%) (Fig. 12.2), of which 236 males (55%) and 191 females (45%). Median age at surgery was 2 years (IQR 0.6–5 years) and median weight was 11 kg (IQR 6–18 kg). The most frequent defect CHDs treated by the use of MS were: VSD (n = 186 patients, 44%), ASD II (n = 145 patients, 34%), p-AVSD (n = 56 patients, 13%), sinus venosus ASDs (n = 13 patients, 3%), PAPVC (n = 12 patients, 2.81%) and others less frequent CHDs (n = 15, 3.5%) (Figs. 12.3–12.5). All the operations were performed by using the CPB and median CPB time was 68 min (IQR 41–96 min). A remote CPB with peripheral cannulation was used in 15 patients (3.5%) (Fig. 12.6). Aortic cross-clamping (ACC) with cardioplegic arrest was used to achieve intra-cardiac repair in 269 patients (63%); median ACC time was 48 min (IQR 36–60 min). The remaining patients were treated by using an IVF (n = 159, 37%), with a median IVF time of 14 min (IQR 9–21 min). Median ICU stay was 2 days (IQR 1–2 days) and median hospitalization was 6 days (IQR 5–7 days).

Forty-eight patients (11.2%) had postoperative complication (Table 12.4).

## RIGHT ANTERIOR MINI-THORACOTOMY

The RAMT was utilized in 349 patients (38.3%) (Fig. 12.7), mainly females (n = 299, 86%). Median age at surgery was 8.3 years (IQR 4–20 years) and median weight was 28 kg (IQR 16–55 kg). The vast majority of patients who were treated by using a RAMT had an ASD II (n = 333 patients, 95.3%). Other CHDs treated with RAMT included: p-AVSD (n = 9 patients, 2.6%), PAPVC (n = 4 patients, 1.1%) and other less frequent CHDs (n = 3 patients, 0.7%) (Figs. 12.3–12.5). All the operations were performed by using the CPB; median CPB time was 35 min (IQR 28–50 min). A remote CPB with peripheral cannulation was used in 94 patients (27%) (Fig. 12.6). IVF was used to achieve intra-cardiac repair in 337 patients (97%), (median IVF time was16 min, IQR 11–25 min). The remaining 12 patients (3.4%) were treated by using ACC with a median ACC time of 35 min (IQR 27–39 min). Median ICU stay was 1 day (IQR 1 day) and median ward stay was 6 days (IQR 5–6 days).

Ten patients (2.9%) had postoperative complications (Table 12.4).

Copyright © 2018 Elsevier Inc. All rights reserved.

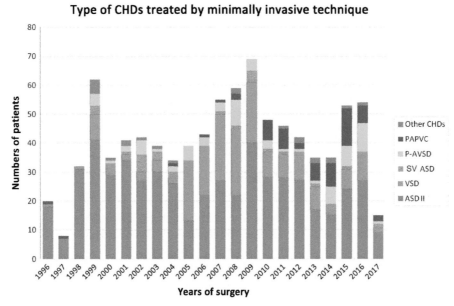

**FIGURE 12.1** **Figure showing the type of CHD treated with minimally invasive approaches by year of surgery.** *ASD II*, Ostium secundum atrial septal defect; *CHD*, congenital heart defects; *PAPVC*, partial anomalous pulmonary venous connection; *p-AVSD*, partial atrio-ventricular septal defects; *SV-ASD*, sino venosus atrial septal defects; *VSD*, ventricular septal defect.

**TABLE 12.1** Right Anterior Mini-Thoracotomy (RAMT) (n = 349 Patients)

| Variables | |
|---|---|
| Males[a] | 50 (14.3) |
| Age at surgery (months)[b] | 100.5 (49.5–241) |
| Weight at surgery (kg)[b] | 27.85 (16–55) |
| Diagnosis[a] | |
| • ASD II | 333 (95.4) |
| • SV ASD | 9 (2.58) |
| • PAPVC | 4 (1.11) |
| • Other CHDs | 3 (0.86) |
| Peripheral cannulation[a] | 94 (26.93) |
| CPB[a] | 345 (100) |
| CPB time (min)[b] | 35 (27.75–50.5) |
| Aortic CC[a] | 12 (3.44) |
| Aortic CC time (min)[b] | 35.5 (27.75–39.5) |
| IVF[a] | 337 (97.12) |
| IVF time (min)[b] | 16 (10.75–25) |
| ICU stay (days)[b] | 1 (1) |
| Hospitalization (days)[b] | 6 (5–6) |
| Complications[a] | 10 (2.87) |

[a]Number of patients and percentage (%).
[b]ACC, Aortic cross-clamping; ASD II, atrial septal defect ostium secundum type; CHDs, congenital heart defects; CPB, cardiopulmonary bypass; ICU, intensive care unit; IQR, median and interquartile range; IVF, induced ventricular fibrillation; PAPVC, partial anomalous pulmonary venous connection; SV ASD, sinus venosus atrial septal defect.

## TABLE 12.2 Mid-Line Mini Sternotomy (MS) (n = 427 Patients)

| Variables | |
| --- | --- |
| Males[a] | 236 (55.3) |
| Age at surgery (months)[b] | 24 (7–60) |
| Weight at surgery (kg)[b] | 11 (6–18) |
| Diagnosis[a] | |
| • VSD | 186 (43.56) |
| • ASD II | 145 (33.96) |
| • p-AVSD | 56 (13.11) |
| • SV ASD | 13 (3.04) |
| • PAPVC | 12 (2.81) |
| • Other less common CHDs | 15 (3.5) |
| Peripheral cannulation[a] | 15 (3.51) |
| CPB[a] | 427 (100) |
| CPB time (min)[b] | 68 (41–96) |
| ACC[a] | 269 (62.99) |
| ACC time (min)[b] | 48 (36–60) |
| IVF[a] | 159 (37.24) |
| IVF time (min)[b] | 14 (9–21) |
| ICU stay (days)[b] | 2 (1–2) |
| Hospitalizations (days)[b] | 6 (5–7) |
| Complications[a] | 48 (11.24) |

[a]Number of patients and percentage (%).
[b]ACC, Aortic cross-clamping; ASD II, atrial septal defect ostium secundum type; CBP, cardiopulmonary bypass; CHDs, congenital heart defects; ICU, intensive care unit; IQR, median and interquartile range; IVF, induced ventricular fibrillation; PAPVC, partial anomalous pulmonary venous connection; SV ASD, sino venosus atrial septal defect.

## TABLE 12.3 Right Lateral Mini-Thoracotomy (RLMT) (n = 134 Patients)

| Variables | |
| --- | --- |
| Males[a] | 74 (55.22) |
| Age at surgery (months)[b] | 115.2 (70–204) |
| Weight at surgery (kg)[b] | 32.25 (20.1–60) |
| Diagnosis[a] | 66 (49.25) |
| • ASD II | 42 (31.34) |
| • PAPVC | 13 (9.70) |
| • p-AVSD | 4 (2.99) |
| • SV ASD | 4 (2.99) |
| • Supra-valvar aortic stenosis | 3 (2.24) |
| • VSD | 2 (1.50) |
| • Other CHDs | |
| Peripheral cannulation[a] | 129 (96.27) |
| CPB[a] | 134 (100) |
| CPB time (min)[b] | 49 (33–80) |
| ACC[a] | 64 (47.76) |
| ACC time (min)[b] | 43 (35–58.5) |
| IVF[a] | 71 (52.99) |
| IVF time (min)[b] | 17 (12–25) |
| ICU stay (days)[b] | 1 (1) |
| Hospitalization (days)[b] | 5 (4–6) |
| Complications[a] | 17 (12.69) |

[a]Number of patients and percentage (%).
[b]ACC, Aortic cross-clamping; ASD II, atrial septal defect ostium secundum type; CHDs, congenital heart defects, cardiopulmonary bypass; ICU, intensive care unit; IQR, median and interquartile range; IVF, induced ventricular fibrillation; PAPVC, partial anomalous pulmonary venous connection; SV ASD, sino venosus septal defect.

**TABLE 12.4 Postoperative Complications (n = 83 in 75 Patients)**

| | MS (427 pts) | RAMT (349pts) | RPMT (134 pts) |
|---|---|---|---|
| Overall | 54 (12.65) | 10 (2.9) | 19 (14.18) |
| Postpericardiotomic syndrome | 14 (3.28) | 3 (0.84) | 2 (1.49) |
| Arrhythmias | 16 (3.75) | 2 (0.58) | 6 (4.48) |
| • transient AV block | 7 (1.64) | 2 (0.58) | 2 (1.49) |
| • permanent AV block | 1 (0.23) | – | – |
| • atrial fibrillation/flutter | 2 (0.47) | – | 2 (1.49) |
| • supraventricular tachycardia | 5 (1.17) | – | – |
| • transient junctional rhythm | 1 (0.23) | – | 2 (1.49) |
| Pneumonia | 4 (0.94) | – | 1 (0.75) |
| Pneumothorax | 3 (0.70) | – | 1 (0.75) |
| Pulmonary hypertension crisis | 3 (0.70) | – | – |
| Pleural effusion | 3 (0.70) | – | 1 (0.75) |
| Mechanical ventilation support >7 days | 5 (1.17) | – | – |
| Reoperation for residual defect | 3 (0.70) | – | – |
| Wound dehiscence | 2 (0.47) | 1 (0.28) | 1 (0.75) |
| LCOS | 1 (0.23) | – | – |
| Bleeding req. surgical revision | – | 2 (0.56) | 1 (0.75) |
| Hemothorax req. surgical revision | – | – | 1 (0.75) |
| Iatrogenic stenosis of femoral artery | – | – | 2 (1.49) |
| Cannulation site hematoma | – | 1 (0.28) | 1 (0.75) |
| Femoral artery dissection | – | 1 (0.28) | – |
| Femoral artero-venous pseudoaneurysm | – | – | 1 (0.75) |
| Femoral vein thrombosis | – | – | 1 (0.75) |

Data presented as number of complications and percentage (%), *AV*, atrio-ventricular; *AV*, atrio-ventricular; *LCOS*, low cardiac output syndrome; *MS*, mini-sternotomy; *RAMT*, right anterior mini-thoracotomy.

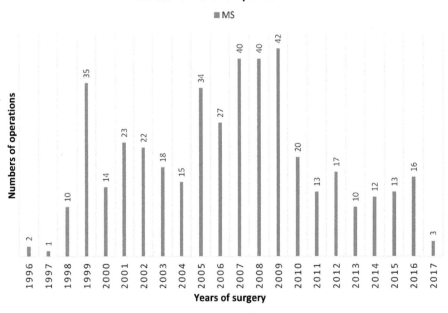

**FIGURE 12.2** Figure showing the number of mid-line mini sternotomies (MS) by year of surgery.

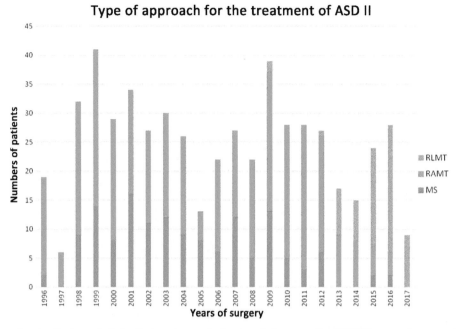

**FIGURE 12.3**   **Figure showing the type of minimally invasive surgical approach for correcting ASD II by year of surgery.** *ASD II*, Ostium secundum atrial septa defects; *MS*, mini-sternotomy; *RAMT*, right anterior mini-thoracotomy; *RLMT*, right-lateral mini-thoracotomy.

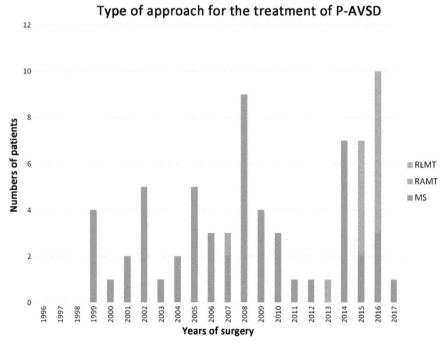

**FIGURE 12.4**   **Figure showing the type of minimally invasive surgical approach for correcting p-AVSDs by year of surgery.** *MS*, Mini-sternotomy; *p-AVSDs*, partial anomalous pulmonary venous connection; partial atrio-ventricular septal defects; *RAMT*, right anterior mini-thoracotomy; *RLMT*, right-lateral mini-thoracotomy.

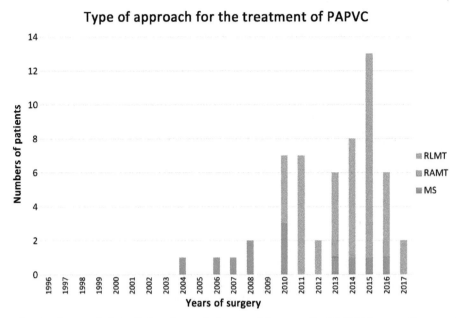

**FIGURE 12.5** **Figure showing the type of minimally invasive surgical approach for correcting PAPVCs by year of surgery.** *MS*, Mini-sternotomy; *PAPVCs*, *RAMT*, right anterior mini-thoracotomy; *RLMT*, right-lateral mini-thoracotomy.

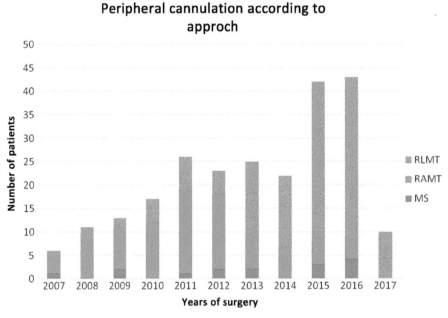

**FIGURE 12.6** **Figure showing the number of patient who underwent a remote CPB with peripheral cannulation by year of surgery (stratified by the type of minimally invasive approach).** *MS*, mini-sternotomy; *RAMT*, right anterior mini-thoracotomy; *RLMT*, right-lateral mini-thoracotomy.

## RIGHT LATERAL (AXILLARY) MINI-THORACOTOMY

The last minimally invasive approach introduced in our institution since 2007, was the right lateral (axillary) mini-thoracotomy (RLMT). It was performed in 134 patients (14.7%) (Fig. 12.8), of which 74 males (55%) and 60 females (45%). Median age at the time of surgery was 9.5 years (IQR 5.8–17 years) and median weight was 32 kg (IQR 20–60 kg).

The most frequent treated CHDs with a RLMT were: ASD II (n = 66, 49%), PAPVC (n = 42, 31%), p-AVSD (n = 13, 10%), p-AVSD (n = 4, 3%), supravalvular aortic stenosis (n = 4, 3%), VSD (n = 3, 2.2%), and others less frequent CHDs (n = 2, 1.8%) (Figs. 12.3–12.5). All the operations were performed by using the CPB; median CPB time was 49 min

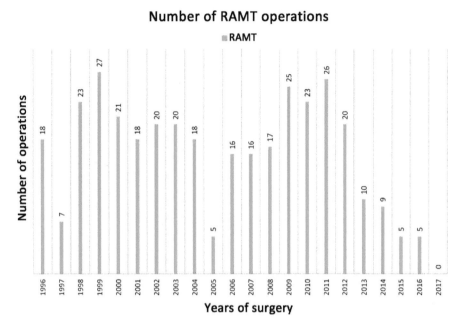

FIGURE 12.7   **Figure showing the number of RAMT by year of surgery.** *RAMT*, Right-anterior mini-thoracotomies.

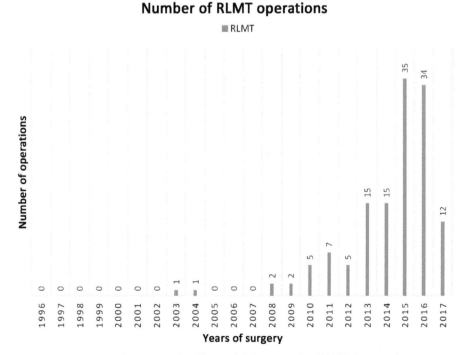

FIGURE 12.8   **Figure showing the number of right-lateral (axillary) mini-thoracotomies (RLMT) by year of surgery.**

(IQR 33–80 min). A remote CPB with peripheral cannulation was used in 129 patients (96%) (Fig. 12.6). IVF was used to achieve intra-cardiac repair in 71 patients (53%); median IVF time was 12 min (IQR 11–25 min). The remaining 64 patients (48%) were treated by using ACC with a median ACC time of 43 min (IQR 35–58 min). Median ICU stay was 1 day (IQR 1–1 day) and median ward stay was 5 days (IQR 4–6 days). Postoperative complications were reported in 17 patients (12.7%) (Table 12.4).

## FURTHER READINGS

[1] Vida VL, Padalino MA, Boccuzzo G, Veshti AA, Speggiorin S, Falasco G, Stellin G. Minimally invasive operation for congenital heart disease: a sex-differentiated approach. J Thorac Cardiovasc Surg 2009;138(4):933–6.

[2] Vida VL, Padalino MA, Motta R, Stellin G. Minimally invasive surgical options in pediatric heart surgery. Expert Rev Cardiovasc Ther 2011;9(6):763–9.

[3] Vida VL, Tessari C, Fabozzo A, Padalino MA, Barzon E, Zucchetta F, Boccuzzo G, Stellin G. The evolution of the right anterolateral thoracotomy technique for correction of atrial septal defects: cosmetic and functional results in prepubescent patients. Ann Thorac Surg 2013;95(1):242–7.

[4] Vida VL, Padalino MA, Boccuzzo G, Stellin G. Near-infrared spectroscopy for monitoring leg perfusion during minimally invasive surgery for patients with congenital heart defects. J Thorac Cardiovasc Surg 2012;143(3):756–7.

[5] Vida VL, Padalino MA, Bhattarai A, Stellin G. Right posterior-lateral mini-thoracotomy access for treating congenital heart disease. Ann Thorac Surg 2011;92(6):2278–80.

# Chapter 13

# Comments and Future Perspectives

Vladimiro L. Vida

*University of Padua, Padua, Italy*

Minimally invasive surgery for congenital heart disease (CHD) refers to therapeutic strategies which are designed to minimize physical trauma associated with surgery [1]. A full sternotomy has been considered for a long time the only surgical access to the heart, with a complete visualization of mediastinal structures [2–6]. Nonetheless, since early 90s, improved surgical results in patients with simple CHD have stimulated the surgeons to adopt minimally invasive techniques, with the aim of combining a good functional outcome with a better cosmetic result [7–11]. These techniques include small incisions, which are designed to reduce operative trauma, with less pain and a shorter rehabilitation time. However, the concern remains about patient safety and proper monitoring during the minimally invasive procedure; in fact a reduced surgical exposure might lead to longer operative times, to a suboptimal treatment and to increased risks [1,12].

Since 1996, we arbitrarily chose different surgical approaches according to patient's gender, keeping in mind patient's safety and satisfaction after the operation [2]. According to our initial protocol, we have adopted the right antero-lateral thoracotomy (RAMT) (Chapter 10) in female patients with atrial septal defects ostium secundum type (ASD II), while a midline mini-sternotomy (MS) was utilized for the correction of simple CHD in males or in females with malformations other than ASDs II. An inferior partial sternotomy ensures a good access to the anterior mediastinum and to the great vessels for central cannulation. It is a safe and effective technique for correction simple CHDs producing comparable functional results to a classic surgical approach. In addition it has been reported that a MS guarantee a better chest stability in the postoperative period that results in a reduction of wound infections and has the theoretical advantages of a shorter hospitalization and also improves the quality of the treatment and the patient's satisfaction [1,2,13].

The use of RAMT has been reported since many years as a valid alternative to a midline sternotomy [5–10,13] and this has been particularly appreciated by the female gender since the surgical incision is located within the mammary sulcus and it is almost always invisible after the operation.

The use of RAMT, in our experience, for correction simple CHDs [2,3,13] has proved to be safe and it offers excellent clinical results, with a very high patient's satisfaction rate (98%) with the cosmetic result of surgery (which increased to 100% in the past 5 years). There was no evidence of scoliosis, restriction to shoulder movement, breast development or lactation problems at follow-up. The reason for a higher satisfaction, mainly in the past decade, is mainly justified by the progressively smaller incisions together with the fact that by the time the incisions were more lateral than in the initial phases of our program. This contributed "per se" to decrease the incidence of less aesthetic medially-located RAMT scars, usually coming off the bra line in the medial portion and consequently also contributed to decrease the possibility to develop an asymmetry between the mammary glands. In addition to the benefit of a limited cutaneous incision, we believe that the quality of our results is related, to an extreme attention that we pay to muscle planes dissection and reconstruction. We strongly suggest that the location of the incision in the submammary area, especially in prepuberal patients, has to be very low, under the right nipple, away from any possibility of future breast development. It is of note that a temporary trivial neuro-sensorial deficit at the mammary area may be present after the operation (in less than 10% of patients) (possibly due to the stretching of the neural fibers during rib retraction) but it will usually disappeared in all patients within 6 months after repair. We found in addition that the length of RAMT incision was also significantly associated to a higher rate of sensitive deficit in the mammary area [2,14].

Conscious of the advantages of lateral minimally invasive approaches (away form the midline of the chest) for correcting simple CHD (technique which was also progressively become the most frequent choses minimally invasive options also chosen by males), we have added to our surgical armamentarium another lateral access, the right lateral mini-thoracotomy

Fundamentals of Congenital Minimally Invasive Cardiac Surgery. http://dx.doi.org/10.1016/B978-0-12-811355-4.00013-7
Copyright © 2018 Elsevier Inc. All rights reserved.

(the so-called "axillary" approach) [15–17]. This technique proves safe and effective, especially in patients where the visualization of the superior vena cava to right atrial junction is mandatory (i.e., patients with anomalous pulmonary venous returns from the right lung to the superior vena cava) and also when the visualization of the aortic root is needed (i.e., aortic cross-clamping or access to the aortic valve or subvalvar area).

The use of induced VF, which was used extensively since early 90s in our institution for correcting patients with ostium scundum atrial septal defect, reveals safe and effective, in particular when in association to the protecting effect of mild systemic hypothermia [2,4,13]. None of the patients where intracardiac repair was achieved by using VF, showed a decreased cardiac function following the procedure or required a postoperative major inotropic support. This technique allows avoiding cumbersome cross-clamping and consequently allows smaller surgical accesses especially in patients requiring RAMT. Nonetheless, we are also aware that a cross-clamping through mini-anterior thoracotomy can be used safely [13], if needed.

A key-point for the evolution of lateral minimally invasive techniques was the use of remote CPB. In fact the use of a peripheral cannulation for CPB establishment [4,18–20] has shown to be a safe and excellent option in selected patients (patients with a body weight above 15 kg). The peripheral cannulation for CPB allows us to further limit the length of our incisions without additional major risks for the patient. In fact, the incidence of complications following peripheral cannulation is very low (1.2% in the initial part of our experience and less than 0.5% during the past 3 years) and none of our patients had acute ischemic limb complications or required the conversion to central cannulation during the procedure. However, a continuous monitoring of the regional tissue oxygenation (a sensor is positioned on the thing of the cannulated leg) by near infrared spectroscopy (NIRS) in patients is mandatory during the phases of the operation. The use in NIRS revealed an interesting tool to continuously check during the CPB any variation of regional oxygen saturation on the cannulated leg, especially in patients with a body weight less than 20 kg, to prevent potential severe vascular complications.

An additional key-point for the evolution of minimally invasive technique technique is represented by the retraction system, which can be easily utilized both in adults and children. This retraction system (Chapter 4), which has been developed for general surgery, has been initially adapted to our patients and employed to improve surgical vision during minimally invasive sternal incision [4]. In these patient it is allow to retract the intact portion of the sternum over the great vessels to facilitate the visualization of the aorta. However, since 2007, it has been also utilized in patients requiring RAMT or right lateral mini-thoracotomy (RLMT), especially when central cannulation or cross clamp are required. In fact, in those patients, aortic cannulation is often the crucial point of the operation and the possibility to have a better exposure of the ascending aorta enhances the safety of the surgical maneuver.

In conclusion, minimally invasive surgery consists in the minimization of surgical accesses with consequent reduction of surgical trauma with consequent less postoperative pain and a more prompt recovery with a consequent reduction of hospital stay and consequently costs. These techniques allow achieving optimal clinical results, which are comparable to the ones of conventional surgery, without increased risks with an additional better cosmetic result and a better perceived quality of the treatment by the patients (higher patient's satisfaction with less psychological discomfort, especially in females, caused by more traditional large scars in the midline of the chest). Since 20 years we are correcting with success simple CHDs and by the time progressively more complex CHD by using minimally invasive technique with optimal functional results and a very high rate of patient's satisfaction. The use of new surgical instruments, the application of vacuum assisted venous drainage, new retracting systems and the use of peripheral CPB allowed us to further minimize over time our surgical approaches without increasing the risk of morbidity for our patients. The more recent lateralization of surgical accesses played in our opinion an additional benefit, which contributes to increase the quality of our results, widening the spectrum of our minimally invasive armamentarium.

Minimally invasive cardiac surgery continues to evolve and expand with growths in technology and surgeon experience. It represents a safe and effective approach for a variety of cardiac surgical diseases and it does not appear to result in differences in short or long-term survival compared with the sternotomy approach. It appears to result in improved pain control, faster recovery to normal activities, decreased length of hospitalization and a very high satisfaction rate, which compare favorably to percutaneous technologies.

Now that a significant amount of data has emerged on the safety and efficacy of minimally invasive cardiac surgery across a range of surgical operations, there is evidence to support the widespread adaptation of such techniques. In the future, there will likely be a greater request for minimally invasive cardiac surgery approaches by patients seeking cardiac surgical options who allow to heal their disease with a faster return to normal activities and improved quality of life.

A multidisciplinary team composed by congenital cardiac surgeons, dedicated cardiac anesthesiologists, intensivists, perfusionists, and nurses is mandatory for the development of a minimally invasive cardiac surgical program for treating CHDs. Minimally invasive cardiac surgery itself will continue to evolve in the future through growing use of percutaneous technology and an ongoing collaborations with interventional cardiologists is mandatory. Continued research is still necessary to assess long-term outcomes of minimally invasive approaches.

# REFERENCES

[1] Bacha E, Kalfa D. Minimally invasive pediatric cardiac surgery. Nat Rev Cardiol 2014;11:24–34.

[2] Vida VL, Padalino MA, Boccuzzo G, Veshti AA, Speggiorin S, Falasco G, et al. Minimally invasive operation for congenital heart disease: a sex-differentiated approach. J Thorac Cardiovasc Surg 2009;138:933–6.

[3] Hagl C, Stock U, Haverich A, Steinhoff G. Evaluation of different minimally invasive techniques in pediatric cardiac surgery: is a full sternotomy always a necessity? Chest 2001;119(2):622–7.

[4] Vida VL, Padalino MA, Motta R, Stellin G. Minimally invasive surgical options in pediatric heart surgery. Expert Rev Cardiovasc Ther 2011;9(6):763–9.

[5] Metras D, Kreitmann B. Correction of cardiac defects through a right thoracotomy in children. J Thorac Cardiovasc Surg 1999;117(5):1040–2.

[6] Lancaster LL, Mavroudis C, Rees AH, Slater AD, Ganzel BL, Gray LA Jr. Surgical approach to atrial septal defect in the female. Right thoracotomy versus sternotomy. Am Surg 1990;56(4):218–21.

[7] Abdel-Rahman U, Wimmer-Greinecker G, Matheis G, Klesius A, Seitz U, Hofstetter R, et al. Correction of simple congenital heart defects in infants and children through a minithoracotomy. Ann Thorac Surg 2001;72(5):1645–9.

[8] De Mulder W, Vanermen H. Repair of atrial septal defects via limited right anterolateral thoracotomy. Acta Chir Belg 2002;102(6):450–4.

[9] Dabritz S, Sachweh J, Walter M, Messmer BJ. Closure of atrial septal defects via a limited right anterolateral thoracotomy as a minimal invasive approach in female patients. Eur J Cardiothor Surg 1999;15:18–23.

[10] Mishaly D, Ghosh P, Preisman S. Minimally invasive congenital cardiac surgery through right anterior minithoracotomy approach. Ann Thorac Surg 2008;85:831–5.

[11] Cremer JT, Boning A, Anssar MB, Kim PY, Pethig K, Harringer W, et al. Different approaches for minimally invasive closure of atrial septal defects. Ann Thorac Surg 1999;67:1648–52.

[12] Bleiziffer S, Schreiber C, Burgkart R, Regenfelder F, Kostolny M, Libera P, et al. The influence of right anterolateral thoracotomy in prepubescent female patients on late breast development and on the incidence of scoliosis. J Thorac Cardiovasc Surg 2004;127(5):1474–80.

[13] Laussen PC, Bichell DP, McGowan FX, Zurakowski D, DeMaso DR, del Nido PJ. Postoperative recovery in children after minimum versus full length sternotomy. Ann Thor Surg 2000;69:591–6.

[14] Vida VL, Tessari C, Fabozzo A, Padalino MA, Barzon E, Zucchetta F, et al. The evolution of right anterior-lateral thoracotomy technique for correction of atrial septal defects: cosmetic and functional results in pre-pubescent patients. Ann Thorac Surg 2013;95(1):242–7.

[15] Vida VL, Padalino MA, Bhattarai A, Stellin G. Right posterior- lateral minithoracotomy access for treating congenital heart disease. Ann Thorac Surg 2011;92(6):2278–80.

[16] Schreiber C, Bleiziffer S, Kostolny M, et al. Minimally invasive midaxillary muscle sparing thoracotomy for atrial septal defect closure in prepubescent patients. Ann Thorac Surg 2005;80:673–6.

[17] Silva Lda F, Silva JP, Turquetto AL, Franchi SM, Cascudo CM, Castro RM, et al. Horizontal right axillary minithoracotomy: aesthetic and effective option for atrial and ventricular septal defect repair in infants and toddlers. Rev Bras Cir Cardiovasc 2014;29(2):123–30.

[18] Vida VL, Tiberio I, Gallo M, Guariento A, Suti E, Pittarello D, et al. Percutaneous internal jugular venous cannulation for extracorporeal circulation during minimally invasive technique in children with congenital heart disease: operative technique and results. Minerva Pediatr 2016;68(5):341–7.

[19] Vida VL, Tessari C, Putzu A, Tiberio I, Guariento A, Gallo M, et al. The peripheral cannulation technique in minimally invasive congenital cardiac surgery. Int J Artif Organs 2016;39(6):300–3.

[20] Vida VL, Padalino MA, Boccuzzo G, Stellin G. Near-infrared spectroscopy for monitoring leg perfusion during minimally invasive surgery for patients with congenital heart defects. J Thorac Cardiovasc Surg 2012;143(3):756–7.

# Index

FBMA LIBRARY
BRITISH MEDICAL ASSOCIATION

WITHDRAWN FROM LIBRARY

CPI Antony Rowe
Chippenham, UK
2018-08-10 22:26